Introduction to Mathematical Logic

Introduction to Mathematical Logic

Agustin Willmott

Larsen & Keller
www.larsen-keller.com

Introduction to Mathematical Logic
Agustin Willmott
ISBN: 978-1-64172-100-4 (Hardback)

Larsen & Keller

Published by Larsen and Keller Education,
5 Penn Plaza,
19th Floor,
New York, NY 10001, USA

Cataloging-in-Publication Data

Introduction to mathematical logic / Agustin Willmott.
    p. cm.
Includes bibliographical references and index.
ISBN 978-1-64172-100-4
1. Logic, Symbolic and mathematical. 2. Mathematics. I. Willmott, Agustin.
QA9 .I58 2019
511.3--dc23

For more information regarding Larsen and Keller Education and its products, please visit the publisher's website www.larsen-keller.com

Table of Contents

Preface

Mathematical logic is a subfield of mathematics that is concerned with the application of formal logic to mathematics. It is closely associated with the foundations of mathematics, metamathematics and theoretical computer science. The study of the deductive power of formal proof systems and the expressive power of formal systems are the unifying themes in mathematical logic. Set theory, recursion theory, proof theory and model theory are the primary subfields in mathematical logic. Each of these fields has a distinct focus. The systems of propositional logic and first-order logic are widely explored for application in the foundations of mathematics. The classical logic systems such as second-order logic or infinitary logic and nonclassical logic systems such as intuitionistic logic are also studied in this field. This book provides comprehensive insights into the field of mathematical logic. It presents the complex subject of mathematical logic in the most comprehensible and easy to understand language. In this book, constant effort has been made to make the understanding of the difficult concepts as easy and informative as possible, for the readers.

A short introduction to every chapter is written below to provide an overview of the content of the book:

Chapter 1, Mathematical logic is an important subfield of mathematics that explores the application of formal logic to mathematics. It can be divided into a number of related fields such as set theory, recursion theory, proof theory and model theory. This chapter has been carefully written to provide an easy understanding of the fundamental concepts of mathematical logic and the scope of its various fields; **Chapter 2**, An understanding of the field of mathematical logic requires an analysis of related theories. These include Gödel's completeness and incompleteness theorem, compactness theorem, Morley's categoricity theorem and Lindström's theorem, among others. This chapter closely examines these important theories for a comprehensive understanding of mathematical logic; **Chapter 3**, A logical system is generally characterized by the properties of consistency, validity, completeness and soundness. The aim of this chapter is to examine the different aspects of logical systems such as first-order, second-order and intuitionistic logic. It also elaborates axiomatic and Hilbert systems; **Chapter 4**, Mathematical logic can be of different forms, like infinitary logic, propositional logic, categorical logic, classical and non-classical logic, algebraic logic, etc. This chapter has been carefully written to provide an easy understanding of the different types of mathematical logic along with their principles and applications; **Chapter 5**, Set theory is a subfield of mathematical logic, which is concerned with the study of sets. All the diverse aspects of set theory have been carefully analyzed in this chapter, such as axioms of set theory, set theory of the continuum, countable and uncountable sets, Zermelo-Fraenkel set theory and Cantor's theorem.

I extend my sincere thanks to the publisher for considering me worthy of this task. Finally, I thank my family for being a source of support and help.

Agustin Willmott

Understanding Mathematical Logic

Mathematical logic is an important subfield of mathematics that explores the application of formal logic to mathematics. It can be divided into a number of related fields such as set theory, recursion theory, proof theory and model theory. This chapter has been carefully written to provide an easy understanding of the fundamental concepts of mathematical logic and the scope of its various fields.

Mathematical logic is best understood as a branch of logic or mathematics. Mathematical logic is often divided into the subfields of model theory, proof theory, set theory and recursion theory. Research in mathematical logic has contributed to, and been motivated by, the study of foundations of mathematics, but mathematical logic also contains areas of pure mathematics not directly related to foundational questions.

One unifying theme in mathematical logic is the study of the expressive power of formal logics and formal proof systems. This power is measured both in terms of what these formal systems are able to prove and in terms of what they are able to define. Thus it can be said that "mathematical logic has become the general study of the logical structure of axiomatic theories."

Earlier names for mathematical logic were symbolic logic (as opposed to philosophical logic) and metamathematics. The former term is still used (as in the Association for Symbolic Logic), but the latter term is now used for certain aspects of proof theory.

Formal Logic

At its core, mathematical logic deals with mathematical concepts expressed using formal logical systems. The system of first-order logic is the most widely studied because of its applicability to foundations of mathematics and because of its desirable properties. Stronger classical logics such as second-order logic or infinitary logic are also studied, along with nonclassical logics such as intuitionistic logic.

Fields of Mathematical Logic

Barwise's "Handbook of Mathematical Logic" (1977) divides mathematical logic into four parts:

- Set theory is the study of sets, which are abstract collections of objects. The basic concepts of set theory such as subset and relative complement are often called naive set theory. Modern research is in the area of axiomatic set theory, which uses logical methods to study which propositions are provable in various formal theories such as Zermelo-Frankel set theory, known as ZFC, or New Foundations set theory, known as NF.

- Proof theory is the study of formal proofs in various logical deduction systems. These proofs are represented as formal mathematical objects, facilitating their analysis by mathematical techniques. Frege worked on mathematical proofs and formalized the notion of a proof.

- Model theory studies the models of various formal theories. The set of all models of a particular theory is called an elementary class. Classical model theory seeks to determine the properties of models in a particular elementary class, or determine whether certain classes of structures form elementary classes. The method of quantifier elimination is used to show that models of particular theories cannot be too complicated.

- Recursion theory, also called computability theory, studies the properties of computable functions and the Turing degrees, which divide the uncomputable functions into sets which have the same level of uncomputability. The field has grown to include the study of generalized computability and definability. In these areas, recursion theory overlaps with proof theory and effective descriptive set theory.

The border lines between these fields, and also between mathematical logic and other fields of mathematics, are not always sharp; for example, Gödel's incompleteness theorem marks not only a milestone in recursion theory *and* proof theory, but has also led to Löb's theorem, which is important in modal logic. The mathematical field of category theory uses many formal axiomatic methods resembling those used in mathematical logic, but category theory is not ordinarily considered a subfield of mathematical logic.

Connections with Computer Science

There are many connections between mathematical logic and computer science. Many early pioneers in computer science, such as Alan Turing, were also mathematicians and logicians.

The study of computability theory in computer science is closely related to the study of computability in mathematical logic. There is a difference of emphasis, however. Computer scientists often focus on concrete programming languages and feasible computability, while researchers in mathematical logic often focus on computability as a theoretical concept and on noncomputability.

The study of programming language semantics is related to model theory, as is program verification (in particular, model checking). The Curry-Howard isomorphism between proofs and programs relates to proof theory; intuitionistic logic and linear logic are significant here. Calculi such as the lambda calculus and combinatory logic are nowadays studied mainly as idealized programming languages.

Computer science also contributes to mathematics by developing techniques for the automatic checking or even finding of proofs, such as automated theorem proving and logic programming.

Groundbreaking Results

- The Löwenheim–Skolem theorem (1919) showed that if a set of sentences in a countable first-order language has an infinite model then it has at least one model of each infinite cardinality.

- Gödel's completeness theorem (1929) established the equivalence between semantic and syntactic definitions of logical consequence in first-order logic.

- Gödel's incompleteness theorems (1931) showed that no sufficiently strong formal system can prove its own consistency.

- The algorithmic unsolvability of the Entscheidungsproblem, established independently by Alan Turing and Alonzo Church in 1936, showed that no computer program can be used to correctly decide whether arbitrary mathematical statements are true.

- The independence of the continuum hypothesis from ZFC showed that an elementary proof or disproof of this hypothesis is impossible. The fact that the continuum hypothesis is consistent with ZFC (if ZFC itself is consistent) was proved by Gödel in 1940. The fact that the negation of the continuum hypothesis is consistent with ZFC (if ZFC is consistent) was proved by Paul Cohen in 1963.

- The algorithmic unsolvability of Hilbert's tenth problem, established by Yuri Matiyasevich in 1970, showed that it is not possible for any computer program to correctly decide whether multivariate polynomials with integer coefficients have any integer roots.

Set Theory

Set theory, branch of mathematics that deals with the properties of well-defined collections of objects, which may or may not be of a mathematical nature, such as numbers or functions. The theory is less valuable in direct application to ordinary experience than as a basis for precise and adaptable terminology for the definition of complex and sophisticated mathematical concepts.

Between the years 1874 and 1897, the German mathematician and logician Georg Cantor created a theory of abstract sets of entities and made it into a mathematical discipline. This theory grew out of his investigations of some concrete problems regarding certain types of infinite sets of real numbers. A set, wrote Cantor, is a collection of definite, distinguishable objects of perception or thought conceived as a whole. The objects are called elements or members of the set.

The theory had the revolutionary aspect of treating infinite sets as mathematical objects that are on an equal footing with those that can be constructed in a finite number of steps. Since antiquity, a majority of mathematicians had carefully avoided the introduction into their arguments of the actual infinite (i.e., of sets containing an infinity of objects conceived as existing simultaneously, at least in thought). Since this attitude persisted until almost the end of the 19th century, Cantor's work was the subject of much criticism to the effect that it dealt with fictions—indeed, that it encroached on the domain of philosophers and violated the principles of religion. Once applications to analysis began to be found, however, attitudes began to change, and by the 1890s Cantor's ideas and results were gaining acceptance. By 1900, set theory was recognized as a distinct branch of mathematics.

Proof Theory

A branch of mathematical logic which deals with the concept of a proof in mathematics and with the applications of this concept in various branches of science and technology.

In the wide meaning of the term, a proof is a manner of justification of the validity of some given assertion. To what extent a proof is convincing will mainly depend on the means employed to substantiate the truth. Thus, in exact sciences certain conditions have been established under which a certain experimental fact may be considered to have been proven (constant reproducibility of the experiment, clear description of the experimental technique, the experimental accuracy, the equipment employed, etc.). In mathematics, where the axiomatic method of study is characteristic, the means of proof were sufficiently precisely established at an early stage of its development. In mathematics, a proof consists of a succession of derivations of assertions from previously derived assertions, and the means of such derivation can be exactly analyzed.

The origin of proof theory can be traced to Antiquity (the deductive method of reasoning in elementary geometry, Aristotelian syllogistics, etc.), but the modern stage in its development begins at the turn of the 19th century with the studies of G. Frege, B. Russell, A.N. Whitehead, E. Zermelo, and, in particular, D. Hilbert. At that time, G. Cantor's research in the theory of sets gave rise to antinomies which cast doubt on the validity of even the simplest considerations concerning arbitrary sets. L.E.J. Brouwer severely criticized certain classical methods of proof of the existence of objects in mathematics, and proposed a radical reconstruction of mathematics in the spirit of intuitionism. Problems concerning the foundations of mathematics became especially timely. Hilbert proposed to separate out the part of practical mathematics known as finitary mathematics, which is unobjectionable as regards both the appearance of antinomies and intuitionistic criticism. Finitary mathematics deals solely with constructive objects such as, say, the natural numbers, and with methods of reasoning that agree with the abstraction of potential realizability but are not concerned with the abstraction of actual infinity. In particular, the use of the law of the excluded middle is restricted. In finitary mathematics, no antinomies have been noted and there is no reason for expecting them to appear. Philosophically, the methods of reasoning of finitary mathematics reflect the constructive processes of real activity much more satisfactorily than those in general set-theoretic mathematics. It was the idea of Hilbert to use the solid ground of finitary mathematics as the foundation of all main branches of classical mathematics. He accordingly presented his formalization method, which is one of the basic methods in proof theory.

In general outline, the formalization method may be described as follows. One formulates a logico-mathematic language (an object language) L in terms of which the assertions of a given mathematical theory T are written as formulas. One then describes a certain class A of formulas of L, known as the axioms of the theory, and describes the derivation rules with the aid of which transitions may be made from given formulas to other formulas. The general term postulates applies to both axioms and derivation rules. The formal theory T^* (a calculus according to a different terminology) is defined by a description of the postulates. Formulas which are obtainable from the axioms of the formal theory by its derivation rules are said to be deducible or provable in that theory. The deduction process itself may be formulated as a derivation tree. T^* is of special interest as regards the contents of the mathematical theory T if the axioms of T^* are records of true statements of T and if the derivation rules lead from true statements to true statements. In such a case T^* may be considered as a precision of a fragment of T, while the concept of a derivation in T^* may be considered as a more precise form of the informal idea of a proof in T, at least within the framework formalized by the calculus T^*. Thus, in constructing the calculus T^*, one must specify, in the first place, which postulates are to be considered suitable from the point of view of the theory T. However, this does not mean that a developed semantics of T must be available at this stage; rather, it is permissible to employ practical habits, to include the most useful or the most theoretically interesting facts among the postulates, etc. The exact nature of the description of derivations in the calculus T^* makes it possible to apply mathematical methods in their study, and thus to give statements on the content and the properties of the theory T.

Proof theory comprises standard methods of formalization of the content of mathematical theories. Axioms and derivation rules of the calculus are usually divided into logical and applied ones. Logical postulates serve to produce statements which are valid by virtue of their form itself, irrespective of the formalized theory. Such postulates define the logic of the formal theory and are formulated in the form of a propositional calculus or predicate calculus. Applied postulates serve to describe truths related to the special features of a given mathematical theory. Examples are: the axiom of choice in axiomatic set theory; the scheme of induction in elementary arithmetic; and bar induction in intuitionistic analysis.

The Hilbert program for the foundations of mathematics may be described as follows. It may be hoped that any mathematical theory T, no matter how involved or how abstract (e.g. the essential parts of set theory), may be formalized as a calculus T^* and that the formulation of the calculus itself requires finitary mathematics alone. Further, by analyzing the conclusions of T^* by purely finitary means, one tries to establish the consistency of T^* and, consequently, to establish the absence of antinomies in T, at least in that part of it which is reflected in the postulates of T^*. It immediately follows, as far as ordinary formalization methods are concerned, that certain very simple statements (in Hilbert's terminology — real statements) are deducible in T^* only if they are true in the finitary sense. It was initially hoped that practically all of classical mathematics could be described in a finitary way, after which its consistency could be demonstrated by finitary means. That this program could not be completed was shown in 1931 by K. Gödel, who proved that, on certain natural assumptions, it is not possible to demonstrate the consistency of T^* even by the powerful tools formalized in T^*. Nevertheless, the study of various formal calculi remains a very important method in the foundations of mathematics. In the first place, it is of interest to construct calculi which reproduce important branches of modern mathematics, with consistency as the guideline, even if it is not yet possible to prove the consistency of such calculi in a manner acceptable to all mathematicians. An example of a calculus of this kind is the Zermelo–Fraenkel system of set theory, in which practically all results of modern set-theoretic mathematics can be deduced. Proofs of non-deducibility, in this theory, of several fundamental hypotheses obtained on the assumption that the theory is consistent indicate that these hypotheses are independent of the set-theoretic principles accepted in mathematics. This in turn may be regarded as a confirmation of the view according to which the existing concepts are insufficient to prove or disprove the hypotheses under consideration. It is in this sense that the independence of Cantor's continuum hypothesis has been established by P. Cohen.

Secondly, extensive study is made of the class of calculi whose consistency can be established by finitary means. Thus Gödel in 1932 proposed a translation converting formulas deducible by classical arithmetical calculus into formulas deducible by intuitionistic arithmetical calculus (i.e. an interpretation of the former calculus into the latter). If the latter is considered consistent (e.g. by virtue of its natural finitary interpretation), it follows that classical arithmetical calculus is self-consistent as well.

Finally, it may be promising to study more extensive methods than Hilbert's traditional finitism which are satisfactory from some other point of view. Thus, remaining in the framework of potential realizability, one may use so-called general inductive definitions. This makes it possible to use semi-formal theories in which some of the derivation rules have an infinite (but constructively generated) set of premises, and to transfer to finitary mathematics many semantic results. This procedure yielded the results obtained by P.S. Novikov (1943), who established the consistency of classical arithmetic using effective functionals of finite type; by C. Spector (1961), who proved the consistency of classical analysis by extending the natural intuitionistic methods of proof to include intuitionistic effective functionals of finite type; and by A.A. Markov (1971), on constructive semantics involving the use of general inductive definitions. In addition, many important problems on calculi may also be considered out of the context of the foundations of mathematics. These include problems of completeness and solvability of formal theories, the problem of independence of certain statements of a given formal theory, etc. It is then unnecessary to impose limitations of definite methods in the reasoning, and it is permissible to develop the theory of proofs as an ordinary mathematical theory, while using any mathematical means of proof that is convincing for the researcher.

A rigorously defined semantics for the formulas of the language under consideration, i.e. a strict definition of the meaning of the statements expressed in that language, serves as an instrument in the study of calculi, and sometimes even as a motivation for the introduction of new calculi. Thus, such a semantics is well-known in classical propositional calculus: Tautologies and only tautologies are deducible in such a calculus. In the general case, in order to prove that a certain formula A is not deducible in a calculus T^* under consideration, it is sufficient to construct a semantics for the formulas of the language of this theory such that all deducible formulas in T^* are true in this semantics, while A is false. A semantics can be classical, intuitionistic or of some other type, depending on the logical postulates it has to agree with. Non-classical semantics are successfully employed in the study of classical calculi — e.g. Cohen's forcing relationship can naturally be regarded as a modification of intuitionistic semantics. In another variant of Cohen's theory, multi-valued semantics are employed; these are models with truth values in a complete Boolean algebra. On the other hand, semantics of the type of Kripke models, defined by a classical set-theoretic method, make it possible to clarify many properties of modal and non-standard logics, including intuitionistic logic.

Proof theory makes extensive use of algebraic methods in the form of model theory. An algebraic system which brings each initial symbol of the language into correspondence with some algebraic objects forms a natural definition of some classical semantics of the language. An algebraic system is said to be a model of the formal theory T^* if all deducible formulas in T^* are true in the semantics generated by the algebraic system. Gödel showed in 1931 that any consistent calculus (with a classical logic) has a model. It was independently shown by A.I. Mal'tsev at a later date that if any finite fragment of a calculus has a model, then the calculus as a whole has a model (the so-called compactness theorem of first-order logic). These two theorems form the foundation of an entire trend in mathematical logic.

A survey of non-standard models of arithmetic established that the concept of the natural numbers is not axiomatizable in the framework of a first-order theory, and that the principle of mathematical induction is independent of the other axioms of arithmetical calculus. The relative nature of the concept of the cardinality of a set in classical mathematics was revealed during the study of countable models for formal theories, the interpretation of which is exclusively based on trivially uncountable models (the so-called Skolem's paradox). Many syntactic results were initially obtained from model-theoretic considerations. In terms of constructions of model theory it is possible to give simple criteria for many concepts of interest to proof theory. Thus, according to Scott's criterion, a class K of algebraic systems of a given language is axiomatizable if and only if it is closed with respect to ultra-products, isomorphisms and taking of elementary subsystems.

A formal theory is said to be decidable if there exists an algorithm which determines for an arbitrary formula A whether or not it is deducible in that theory. It is known that no formal theory containing a certain fragment of the theory of recursive functions is decidable. It follows that elementary arithmetic, the Zermelo–Fraenkel system and many other theories are undecidable as well. Proof theory disposes of powerful methods for the interpretation of theories in terms of other theories; such interpretations may also be used to establish the undecidability of several very simple calculi in which recursive calculi are not interpreted directly. Examples are elementary group theory, the theory of two equivalence relations, an elementary theory of fractional order, etc. On the other hand, examples are available of interesting decidable theories such as elementary geometry, the elementary theory of real numbers, and the theory of sets of natural numbers with a unique successor operation. The decidability of a theory is demonstrated by model-theoretic and syntactic methods. Syntactic methods often yield simpler decidability algorithms. For instance, the decidability of the elementary theory of P-adic numbers was first established by model-theoretic methods. Subsequently a primitive-recursive algorithm was found for the recognition of decidability of this theory by a certain modification of the syntactic method of quantifier elimination. Estimates of the complexity of decidability algorithms of theories are of importance. As a rule, a primitive-recursive solution algorithm is available for decidable theories, and the problem is to find more exact complexity bounds. A promising direction in such studies is the decidability of real fragments of known formal theories. In this connection classical predicate calculus has been studied in much detail, where an effective description has been given of all decidable and undecidable classes of formulas, in terms of the position of quantifiers in the formula and the form of the predicate symbols appearing in the formula. A number of decidable fragments of arithmetical calculus and of elementary set theory have been described.

Methods for estimating the complexity of derivations have attracted the attention of researchers. This area of research comprises problems such as finding relatively short formulas that are derivable in a complex manner, or formulas yielding a large number of results in a relatively simple manner. Such formulas must be regarded as expressing the "depth" of the facts in the theory. Natural measures of the complexity of a proof are studied: the length of the proof; the time needed to find a solution; the complexity of the formulas used in the proof, etc. This is the domain of contact between proof theory and the methods of theoretical cybernetics.

Constructive Mathematics

Constructive mathematics is mathematics done without the principle of excluded middle, or other principles, such as the full axiom of choice, that imply it, hence without "non-constructive" methods of formal proof, such as proof by contradiction. This is in contrast to *classical mathematics*, where such principles are taken to hold.

Mathematics built up in connection with a certain constructive mathematical view on the world that usually seeks to relate statements on the existence of mathematical objects with the possibility of their construction, rejecting thereby a number of standpoints of traditional set-theoretic mathematics and leading to the appearance of pure existence theorems (in particular, the abstraction of actual infinity and the rejection of the universal nature of the law of the excluded middle). The constructive trend in mathematics has emerged in some form or other throughout its history, although it appears to be C.F. Gauss who first stated explicitly the difference, being the principal one in constructive mathematics, between potential infinity and the actual mathematical infinity; he objected to the use of the latter. Subsequent critical steps in this direction were taken by L. Kronecker, H. Poincaré and especially L.E.J. Brouwer. Brouwer's criticism, written at the time of the crisis in the foundations of mathematics at the beginning of the 20th century, vigorously renounced both the belief in the existential character of infinite sets and the conviction in the admissibility of unrestricted extrapolation of classical logical principles, especially of the law of the excluded middle. As an alternative set-theoretic approach, Brouwer, and subsequently his followers, developed an original program for constructing mathematics, known nowadays under the name of intuitionism. Brouwer's intuitionistic mathematics can be regarded as the first systematic attempt at building up mathematics on a constructive basis. In parallel with the success of the intuitionists there appeared in proof theory (created by D. Hilbert with the aim of justifying set-theoretic mathematics) a number of original ideals that served as a starting point for constructive trends differing from intuitionism. A significant number of papers on this topic (and here a fairly broad spectrum of interpretations of terms like "constructive", "effective", etc., by various researchers can be found) have depended on the success achieved in the study of the mathematical concept of algorithm (again, under the influence of the ideas of Hilbert). One of the most generally followed and definitive approaches to the development of constructive mathematics on this basis was provided by A.A. Markov of the Soviet school of constructive mathematics, the formation of the fundamental concepts of which go back to the 1950's. The term "constructive mathematics" itself (the constructive trend in mathematics) is often applied in a more general sense of the word, referring to this Soviet approach to constructive mathematics. In what follows this term will be used in this latter sense.

Constructive mathematics can be briefly characterized by the following basic features: 1) the objects of study are constructive processes and the constructive objects that arise as a result of carrying out these processes; 2) the examination of the constructive processes and objects is carried out within the framework of the abstraction of potential realizability with total exclusion of the idea of an actual infinity; 3) the intuitive concept of effectiveness is connected with a precise notion of algorithm; and 4) a special constructive logic is used that takes into account the specifics of the constructive processes and objects.

The notions of constructive process and constructive object are primitive; the ideas concerning them have practical material human activity as their basis. Examples of constructive processes are assembling watches on a conveyor belt, the complete or partial dismantling of them in the repair shop, the typesetting of texts (with corrections) in printing works, the formation and decomposition of railway trains, etc. The characteristic feature of constructive processes is an operation passing through separate stages in the context of certain clearly indicated rules with elementary objects that are clearly distinguishable from one another and that are considered as indecomposable in the course of these processes. The resulting configuration, composed of the original elementary objects, are regarded as constructive objects. Constructive mathematics does not have the need to delve into the general notion of a constructive process or object since one special form of the constructive objects proves to be entirely adequate for its needs, namely a word over some alphabet or other.

The examination of words (this concept is also supposed to be primitive) takes place as follows.

To begin with, a certain alphabet is fixed, that is, a list of indecomposable, demonstrably different elementary signs (letters). Each letter of the alphabet can be copied; linear strings or symbols arising from sequences of acts of such copying are regarded as words over the original alphabet. It is convenient to add to the words over the given alphabet the empty word, that is, a string containing no symbol. For example, the string 5): "abbbccd" , and 6): "book" , are words over the English alphabet. In its handling of words, constructive mathematics — and here its abstract nature becomes apparent — uses abstraction by identification and potential realization. The first of these enables one, by abstracting from different copies and the original, to speak about different copies of a given letter as well as about it being one and the same letter. For example, in the word 5) the letter "b" of the English alphabet occurs three times, whereas in reality, three distinct copies of the original letter were reproduced in writing the word down. This agreement naturally extends to words that are identically written (but graphically distinct). For example, one speaks of the two concrete words, the word "book" and the word 6), as one and the same word. On the assumption of abstraction by identification there emerges the primitive ability of the human being to "scan" words, that is, to be able repeatedly and consistently to recognize strings of symbols as being the same or different. Hilbert pointed out this ability as a minimal premise for any scientific activity. Abstraction of potential realizability enables one to ignore in discussions concerning the description of words the real-life constraints of space, time and material. In this way one can begin to argue about imagined very long words as though they really existed, in particular, it is regarded as possible to write on the right (or left) of a given word any other word. This implies the possibility of considering arbitrary large natural numbers as well as adding any two natural numbers, since natural numbers can be regarded, for example, as words of the form $0, 0|, 0| |$, etc. over the alphabet $0|$. Apart from this, the abstraction of potential realizability does not enable one to regard "infinite" words and the collection of "all" words over a given alphabet as complete in their own right (in particular, the natural series is not regarded as a complete object). Considerations of this sort require drawing upon a stronger abstraction — the abstraction of actual infinity, which constructive mathematics rejects.

The acceptance of the abstraction of potential realizability leads to the fact that not only elementary, completely visible constructive processes are considered (for example the writing down of short words), but also conceptual constructive processes which are not subject to actual reproduction. Such processes are defined by their instructions; these instructions themselves then become an object of study. The instruction giving the constructive process (for simplicity, one is dealing with processes operating on words) has to be understood by all and must perfectly uniquely determine from stage to stage the sequential construction of words; furthermore, the stages must be elementary, that is, they presuppose nothing other than the ability of the reader to read, write (or erase) words. These stages thus reduce to the writing down and graphical comparison of certain words, as well as to the substitution of occurrences of single words into other alternative words. The termination of the process is defined by the instruction itself and may depend on results obtained at stages prior to the final one; here the acceptance of a decision concerning the conclusive nature of a given stage must also bear the elementary character just described. It is possible that no stage proves to be conclusive, that is, after each complete stage, the given instruction requires that the following stage be completed. To such an instruction one cannot associate any potentially implementable constructive process.

Here it turns out to be convenient to use conventional terminology, according to which the corresponding instruction determines an unboundedly extendable (potentially infinite) process. For the justification of this terminology, one could also extend the original idea on constructive processes by considering along with potentially realizable processes more abstract formations, such as processes identified with their instructions. In connection with the appearance of the potential infinity of constructive processes the question arises as to the means by which one can be certain that a constructive process defined by a given instruction terminates. Constructive mathematics here applies an important principle, called the constructive selection principle (Markov's principle), enabling one to establish such facts by the method of contradiction, that is, the reduction to absurdity of the hypothesis that the corresponding constructive process is potentially infinite. Examples of instructions: 7): write $|$; 8): write on the right of an arbitrary word over the alphabet o $|$ the word $|$; 9): 9_a): write $|$ and go to 9_b); 9_b): erase $|$ (that is, replace this letter by the empty word) and go to 9_a); 10): 10_a): add to the right of an arbitrary word over the alphabet o $|$ and go to 10_b); 10_b): if the currently processed word is o $|$ $|$, then terminate the process, otherwise return to 10_a); 11): 11_a): write o and go to 11); 11_b): add $|$ to the right of the currently processed word and go to 11_c); 11_c): if a perfect natural number has been obtained, then terminate the process, otherwise add $|$ to the currently processed word and return to 11_b).

The instruction 7) defines a constructive process that terminates in one stage by writing the one-letter word $|$. The process of carrying out 9) is potentially infinite. It is not known whether the constructive process defined by 11) terminates (in 11) ideas from number theory have been used for brevity; clearly, a longer instruction of this sort is possible using exclusively the ability to scan, write and compare words over the alphabet o). The instructions 8) and 10) have a rather special character: their accomplishment can begin with any word over the indicated alphabet, and, furthermore, the constructive process defined by 8) always terminates, while at the same time, as in the case of 10), it is potentially infinite for certain initial words. Instructions of the above types are conventionally called algorithms (in the present context, one is concerned with algorithms operating on words).

The constructive treatment of existential statements leads to the necessity of considering algorithms. The assertion that a constructive objects with a given property exists, that is, a statement of the form 12): $\exists x A(x)$, in relation to ideas about constructive objects as the results of constructive processes, is regarded as established in constructive mathematics only in the case when a potentially realizable constructive process can be indicated which terminates by the construction of the required object. In a similar manner, establishing an existential statement with a parameter 13):

$\forall x \exists y \, A(x,y)$ ("for all x there exists an y such that Ax, y") presupposes that one can indicate a "general" constructive process starting from an arbitrary constructive object x of a given primitive type and terminating with the construction of a required y. In other words, 13) expresses the existence of an algorithm for finding y starting from x. Such a treatment of existence entails a constructive interpretation of disjunction: The statement $A \vee B$ ("A or B") is considered to be established only if a constructive process can be presented that terminates with the indication that one of its components is true. The further explanation of the meaning of statements of a more complex structure and the elaboration of rules for handling them in accordance with the original constructive aims, constitutes the problem of constructive semantics and constructive logic. The constructive treatment of existential and disjunctive statements given above is essentially different from the traditional one: In set-theoretic mathematics, for example, 12) can be proved by reducing its denial to absurdity. Such a proof usually contains no method for constructing the required constructive object. Constructive mathematics maintains that such an argument does not prove 12) but only its "double negationdouble negation" , that is, $\neg\neg \exists x A(x)$. The latter statement is generally regarded in constructive mathematics as being weaker than 12). Thus, constructive mathematics does not apply the rule of cancelling the double negation nor, consequently, the law of the excluded middle (the constructive treatment of disjunction also indicates that there is no basis for accepting the latter).

The primitive mathematical structures of natural, integer and rational numbers can be directly treated as words of certain simple types over a fixed alphabet; here the relations of equality and order are easily reduced to graphical coincidence and difference of words. The introduction of more complex structures such as the real numbers and functions on them, etc., are realized in constructive mathematics by the notion of algorithm, which, roughly speaking, here plays the same role as the concept of a function in traditional mathematics. In considering the intuitive notions of algorithm too vague for such constructions, constructive mathematics makes here an important step by standardizing the usable algorithms by means of accepting one of the modern precise definitions of this notion together with a corresponding hypothesis in the form of the Church thesis, which asserts that the operational possibilities achievable by algorithms are the same in the intuitive and in the precise senses of the word. In fact, the widest application in constructive mathematics has been attained by Markov's normal algorithms. The constructive treatment of existence also leads to the need of making precise the notion of algorithm. For example, the negation of 13) is the assertion of the impossibility of there being some algorithm, while intuitive ideas that suffice for recognizing some or other concrete instruction as an algorithm, in principle do not allow one to obtain some non-trivial theorems on impossibility. On the basis of the principle set out above and using the modern theory of algorithms in constructive mathematics, one develops a number of mathematical disciplines including constructive mathematical analysis, elements of functional analysis, differential equations, the theory of functions of a complex variable, etc. The theoretical models obtained in this way are based on more modest abstractions than the usual system and although they are not so transparent and elegant as the traditional methods, they are nevertheless, it would seem, capable of serving the same circle of applications.

Having a common critical source with the intuitionistic mathematics of Brouwer and borrowing from it a number of constructions and ideas, constructive mathematics has a certain similarity to the latter. Apart from this, there is also a principal difference both of a general philosophical and a concrete mathematical nature. In the first place, constructive mathematics does not single out the belief, peculiar to intuitionism, in the primitive character of mathematical intuition by supporting that this intuition itself is formed under the influence of practical activities of the human being. Accordingly, abstraction in constructive mathematics proceeds not from mental constructions as in intuitionism, but from the simplest real observable constructive processes. In the mathematical setting, constructive mathematics does not apply the deductive concept of free choice sequences, which goes beyond the framework of constructive processes and objects, nor the intuitionistic theory of the continuum as a medium of free formation based on them. On the other hand, intuitionistic mathematics does not accept the constructive selection principle and does not regard it necessary to eliminate intuitive algorithms in favour of corresponding precise definitions. It should be pointed out that in a recent years a certain trend has emerged towards bringing the constructive and intuitionistic approaches closer together: in certain constructive investigations, in particular those relating to semantics, inductive definitions are used and corresponding to them, inductive proofs recalling Brouwer's construction in the proof of his so-called bar-theorem which occupies one of the central places in intuitionistic mathematics.

Model Theory

Model theory is the study of the relationship between the *language* of mathematics and the actual world of mathematical objects and structures. It is one of the most active areas of logic and has many connections with other areas of mathematics.

Model theory is a different branch of mathematical logic that concerns itself with the semantics and syntactics of abstract mathematical structures, such as the field of complex numbers $(C, +, \cdot)$ and linear orders such as $(\mathbb{Q}, <)$. One of the results that modernized the subject was Morley's categoricity theorem. A fundamental result in linear algebra is that any two vector spaces that have the same dimension and same field of scalars are isomorphic. If the field of scalars is the rational numbers, then the dimension of an uncountable vector space is the same as its cardinality. Thus, any two vector spaces over $\mathbb{Q}$, which have the same uncountable cardinality, are in fact isomorphic. The categoricity theorem is a vast generalization of this; it asserts that, for any theory T and uncountable cardinals κ and λ, if any two models of T of cardinality κ are isomorphic, then any two models of cardinality λ are isomorphic. The proof proceeds by showing that, even in this generality, theories that satisfy the hypothesis of the theorem admit an abstract notion of a basis, much as in the setting of vector spaces.

This rather extreme characteristic of a theory is by no means typical. For example, both $\mathbb{R}$ and $\mathbb{Q} \cup (0, 1)$ are dense linear orders of the same cardinality, but they are not isomorphic: every interval in $\mathbb{R}$ is uncountable, whereas the interval of points between 2 and 3 in $\mathbb{Q} \cup (0, 1)$ is countable. Theories that are close to that of the algebraic structure $(C, +, \cdot)$ are generally regarded in model theory as being tame, whereas those similar to the dense linear order $(\mathbb{R}, <)$ are regarded as being wild. In the 1970s, Shelah initiated his classification program in model theory in an attempt to separate the tame from the wild and to stratify what lay in between.

One method of stratifying theories is Keisler's order, which measures how easily the models of the theory become saturated by taking ultrapowers. The ultrapower construction is a useful tool by which a structure is enlarged while preserving its theory. The logical properties of the ultrapower are obtained by integration via a certain binary valued measure, known as an ultrafilter. Such enlargements have a greater tendency to contain solutions to large systems of logical equations. Those structures that have solutions to any consistent system of logical equations, which are smaller than the structure itself, are said to be saturated. The tamer the theory, the more likely its ultrapowers are to be saturated.

It was long known that the theories of $(C, +, \cdot)$ and $(\mathbb{R}, <)$ provided examples at the opposite ends of this spectrum, and that the presence of a definable ordering in the structure was an indication of high complexity. Shelah also characterized, in model theoretic terms, the smallest two classes in Keisler's order: these classes together are the "stable theories". No model theoretic characterization was known, however, of the maximal theories in Keisler's order. Over the decades that followed, very little progress was made in this direction.

Branches of Model Theory

This topic focuses on finitary first order model theory of infinite structures. Finite model theory, which concentrates on finite structures, diverges significantly from the study of infinite structures in both the problems studied and the techniques used. Model theory in higher-order logics or infinitary logics is hampered by the fact that completeness and compactness do not in general hold for these logics. However, a great deal of study has also been done in such logics.

Informally, model theory can be divided into classical model theory, model theory applied to groups and fields, and geometric model theory. A missing subdivision is computable model theory, but this can arguably be viewed as an independent subfield of logic.

Examples of early theorems from classical model theory include Gödel's completeness theorem, the upward and downward Löwenheim–Skolem theorems, Vaught's two-cardinal theorem, Scott's isomorphism theorem, the omitting types theorem, and the Ryll-Nardzewski theorem. Examples of early results from model theory applied to fields are Tarski's elimination of quantifiers for real closed fields, Ax's theorem on pseudo-finite fields, and Robinson's development of non-standard analysis. An important step in the evolution of classical model theory occurred with the birth of stability theory (through Morley's theorem on uncountably categorical theories and Shelah's classification program), which developed a calculus of independence and rank based on syntactical conditions satisfied by theories.

During the last several decades applied model theory has repeatedly merged with the more pure stability theory. The result of this synthesis is called geometric model theory in this topic (which is taken to include o-minimality, for example, as well as classical geometric stability theory). An example of a theorem from geometric model theory is Hrushovski's proof of the Mordell–Lang conjecture for function fields. The ambition of geometric model theory is to provide a *geography of mathematics* by embarking on a detailed study of definable sets in various mathematical structures, aided by the substantial tools developed in the study of pure model theory.

Universal Algebra

Fundamental concepts in universal algebra are signatures σ and σ-algebras. Since these concepts are formally defined in the article on structures, the present topic is an informal introduction which consists of examples of the way these terms are used.

The standard signature of rings is $\sigma_{ring} = \{\times,+,-,0,1\}$, where $\times$ and $+$ are binary, $-$ is unary, and 0 and 1 are nullary.

The standard signature of semirings is $\sigma_{smr} = \{\times,+,0,1\}$, where the arities are as above.

The standard signature of groups (with multiplicative notation) is $\sigma_{grp} = \{\times,^{-1},1\}$, where $\times$ is binary, $^{-1}$ is unary and 1 is nullary.

The standard signature of monoids is $\sigma_{mnd} = \{\times,1\}$.

A ring is a σ_{ring}-structure which satisfies the identities $u + (v + w) = (u + v) + w$, $u + v = v + u$, $u + 0 = u$, $u + (-u) = 0$, $u \times (v \times w) = (u \times v) \times w$, $u \times 1 = u$, $1 \times u = u$, $u \times (v + w) = (u \times v) + (u \times w)$ and $(v + w) \times u = (v \times u) + (w \times u)$.

A group is a σ_{grp}-structure which satisfies the identities $u \times (v \times w) = (u \times v) \times w$, $u \times 1 = u$, $1 \times u = u$, $u \times u^{-1} = 1$ and $u^{-1} \times u = 1$.

A monoid is a σ_{mnd}-structure which satisfies the identities $u \times (v \times w) = (u \times v) \times w$, $u \times 1 = u$ and $1 \times u = u$.

A semigroup is a $\{\times\}$-structure which satisfies the identity $u \times (v \times w) = (u \times v) \times w$.

A magma is just a $\{\times\}$-structure.

This is a very efficient way to define most classes of algebraic structures, because there is also the concept of σ-homomorphism, which correctly specializes to the usual notions of homomorphism for groups, semigroups, magmas and rings. For this to work, the signature must be chosen well.

Terms such as the σ_{ring}-term $t(u,v,w)$ given by $(u + (v \times w)) + (-1)$ are used to define identities $t = t'$, but also to construct free algebras. An equational class is a class of structures which, like the examples above and many others, is defined as the class of all σ-structures which satisfy a certain set of identities. Birkhoff's theorem states:

A class of σ-structures is an equational class if and only if it is not empty and closed under subalgebras, homomorphic images, and direct products.

An important non-trivial tool in universal algebra are ultraproducts $\Pi_{i \in I} A_i / U$, where I is an infinite set indexing a system of σ-structures A_i, and U is an ultrafilter on I.

While model theory is generally considered a part of mathematical logic, universal algebra, which grew out of Alfred North Whitehead's (1898) work on abstract algebra, is part of algebra. This is reflected by their respective MSC classifications. Nevertheless, model theory can be seen as an extension of universal algebra.

Finite Model Theory

Finite model theory is the area of model theory which has the closest ties to universal algebra. Like some parts of universal algebra, and in contrast with the other areas of model theory, it is mainly concerned with finite algebras, or more generally, with finite σ-structures for signatures σ which may contain relation symbols as in the following example:

The standard signature for graphs is $\sigma_{grph}=\{E\}$, where E is a binary relation symbol.

A graph is a σ_{grph}-structure satisfying the sentence $\forall u \forall v (uEv \rightarrow vEu)$.

A σ-homomorphism is a map that commutes with the operations and preserves the relations in σ. This definition gives rise to the usual notion of graph homomorphism, which has the interesting property that a bijective homomorphism need not be invertible. Structures are also a part of universal algebra; after all, some algebraic structures such as ordered groups have a binary relation <. What distinguishes finite model theory from universal algebra is its use of more general logical sentences (as in the example above) in place of identities. (In a model-theoretic context an identity $t=t'$ is written as a sentence $\forall u_1 u_2 \ldots u_n (t = t')$.)

The logics employed in finite model theory are often substantially more expressive than first-order logic, the standard logic for model theory of infinite structures.

First-order Logic

Whereas universal algebra provides the semantics for a signature, logic provides the syntax. With terms, identities and quasi-identities, even universal algebra has some limited syntactic tools; first-order logic is the result of making quantification explicit and adding negation into the picture.

A first-order formula is built out of atomic formulas such as $R(f(x,y),z)$ or $y = x + 1$ by means of the Boolean connectives $\neg, \wedge, \vee, \rightarrow$ and prefixing of quantifiers $\forall v$ or $\exists v$. A sentence is a formula in which each occurrence of a variable is in the scope of a corresponding quantifier. Examples for formulas are φ (or φ(x) to mark the fact that at most x is an unbound variable in φ) and ψ defined as follows:

$$\varphi = \forall u \forall v (\exists w(x \times w = u \times v) \rightarrow (\exists w(x \times w = u) \vee \exists w(x \times w = v))) \wedge x \neq 0 \wedge x \neq 1,$$

$$\psi = \forall u \forall v ((u \times v = x) \rightarrow (u = x) \vee (v = x)) \wedge x \neq 0 \wedge x \neq 1.$$

(Note that the equality symbol has a double meaning here.) It is intuitively clear how to translate such formulas into mathematical meaning. In the σ_{smr}-structure $\mathcal{N}$ of the natural numbers, for example, an element n satisfies the formula φ if and only if n is a prime number. The formula ψ similarly defines irreducibility. Tarski gave a rigorous definition, sometimes called "Tarski's definition of truth", for the satisfaction relation $\models$, so that one easily proves:

$\mathcal{N} \models \varphi(n) \Leftrightarrow n$ is a prime number.

$\mathcal{N} \models \psi(n) \Leftrightarrow n$ is irreducible.

A set T of sentences is called a (first-order) theory. A theory is satisfiable if it has a model $\mathcal{M} \models T$, i.e. a structure (of the appropriate signature) which satisfies all the sentences in the set T. Consistency of a theory is usually defined in a syntactical way, but in first-order logic by the completeness theorem there is no need to distinguish between satisfiability and consistency. Therefore, model theorists often use "consistent" as a synonym for "satisfiable".

A theory is called categorical if it determines a structure up to isomorphism, but it turns out that this definition is not useful, due to serious restrictions in the expressivity of first-order logic. The Löwenheim–Skolem theorem implies that for every theory T having a countable signature which has an infinite model for some infinite cardinal number, then it has a model of size κ for any infinite cardinal number κ. Since two models of different sizes cannot possibly be isomorphic, only finitary structures can be described by a categorical theory.

Lack of expressivity (when compared to higher logics such as second-order logic) has its advantages, though. For model theorists, the Löwenheim–Skolem theorem is an important practical tool rather than the source of Skolem's paradox. In a certain sense made precise by Lindström's theorem, first-order logic is the most expressive logic for which both the Löwenheim–Skolem theorem and the compactness theorem hold.

As a corollary (i.e., its contrapositive), the compactness theorem says that every unsatisfiable first-order theory has a finite unsatisfiable subset. This theorem is of central importance in infinite model theory, where the words "by compactness" are commonplace. One way to prove it is by means of ultraproducts. An alternative proof uses the completeness theorem, which is otherwise reduced to a marginal role in most of modern model theory.

Axiomatizability, Elimination of Quantifiers and Model-completeness

The first step, often trivial, for applying the methods of model theory to a class of mathematical objects such as groups, or trees in the sense of graph theory, is to choose a signature σ and represent the objects as σ-structures. The next step is to show that the class is an elementary class, i.e. axiomatizable in first-order logic (i.e. there is a theory T such that a σ-structure is in the class if and only if it satisfies T). E.g. this step fails for the trees, since connectedness cannot be expressed in first-order logic. Axiomatizability ensures that model theory can speak about the right objects. Quantifier elimination can be seen as a condition which ensures that model theory does not say too much about the objects.

A theory T has quantifier elimination if every first-order formula $\varphi(x_1,...,x_n)$ over its signature is equivalent modulo T to a first-order formula $\psi(x_1,...,x_n)$ without quantifiers, i.e. $\forall x_1 ... \forall x_n (\phi(x_1,...,x_n) \leftrightarrow \psi(x_1,...,x_n))$ holds in all models of T. For example, the theory of algebraically closed fields in the signature $\sigma_{ring}=(\times,+,-,0,1)$ has quantifier elimination because every formula is equivalent to a Boolean combination of equations between polynomials.

A substructure of a σ-structure is a subset of its domain, closed under all functions in its signature σ, which is regarded as a σ-structure by restricting all functions and relations in σ to the subset. An embedding of a σ-structure A into another σ-structure B is a map f: A → B between the domains which can be written as an isomorphism of A with a substructure of B . Every embedding is an injective homomorphism, but the converse holds only if the signature contains no relation symbols.

If a theory does not have quantifier elimination, one can add additional symbols to its signature so that it does. Early model theory spent much effort on proving axiomatizability and quantifier elimination results for specific theories, especially in algebra. But often instead of quantifier elimination a weaker property suffices:

A theory T is called model-complete if every substructure of a model of T which is itself a model of

T is an elementary substructure. There is a useful criterion for testing whether a substructure is an elementary substructure, called the Tarski–Vaught test. It follows from this criterion that a theory T is model-complete if and only if every first-order formula $\varphi(x_1,...,x_n)$ over its signature is equivalent modulo T to an existential first-order formula, i.e. a formula of the following form:

$$\exists v_1 ... \exists v_m \psi(x_1,...,x_n,v_1,...,v_m),$$

where ψ is quantifier free. A theory that is not model-complete may or may not have a model completion, which is a related model-complete theory that is not, in general, an extension of the original theory. A more general notion is that of model companions.

Categoricity

As observed in this topic on first-order logic, first-order theories cannot be categorical, i.e. they cannot describe a unique model up to isomorphism, unless that model is finite. But two famous model-theoretic theorems deal with the weaker notion of κ-categoricity for a cardinal κ. A theory T is called κ-categorical if any two models of T that are of cardinality κ are isomorphic. It turns out that the question of κ-categoricity depends critically on whether κ is bigger than the cardinality of the language (i.e. $\aleph_0 + |\sigma|$, where $|\sigma|$ is the cardinality of the signature). For finite or countable signatures this means that there is a fundamental difference between $\aleph_0$-cardinality and κ-cardinality for uncountable κ.

A few characterizations of $\aleph_0$-categoricity include:

For a complete first-order theory T in a finite or countable signature the following conditions are equivalent:

1. T is $\aleph_0$-categorical.

2. For every natural number n, the Stone space $S_n(T)$ is finite.

3. For every natural number n, the number of formulas $\varphi(x_1, ..., x_n)$ in n free variables, up to equivalence modulo T, is finite.

This result, due independently to Engeler, Ryll-Nardzewski and Svenonius, is sometimes referred to as the Ryll-Nardzewski theorem.

Further, $\aleph_0$-categorical theories and their countable models have strong ties with oligomorphic groups. They are often constructed as Fraïssé limits.

Michael Morley's highly non-trivial result that (for countable languages) there is only *one* notion of uncountable categoricity was the starting point for modern model theory, and in particular classification theory and stability theory:

> Morley's categoricity theorem
>
> If a first-order theory T in a finite or countable signature is κ-categorical for some uncountable cardinal κ, then T is κ-categorical for all uncountable cardinals κ.

Uncountably categorical (i.e. κ-categorical for all uncountable cardinals κ) theories are from many points of view the most well-behaved theories. A theory that is both $\aleph_0$-categorical and un-countably categorical is called totally categorical.

Other Basic Notions

Reducts and Expansions

A field or a vector space can be regarded as a (commutative) group by simply ignoring some of its structure. The corresponding notion in model theory is that of a reduct of a structure to a subset of the original signature. The opposite relation is called an *expansion* - e.g. the (additive) group of the rational numbers, regarded as a structure in the signature {+,0} can be expanded to a field with the signature {×,+,1,0} or to an ordered group with the signature {+,0,<}.

Similarly, if σ' is a signature that extends another signature σ, then a complete σ'-theory can be restricted to σ by intersecting the set of its sentences with the set of σ-formulas. Conversely, a complete σ-theory can be regarded as a σ'-theory, and one can extend it (in more than one way) to a complete σ'-theory. The terms reduct and expansion are sometimes applied to this relation as well.

Interpretability

Given a mathematical structure, there are very often associated structures which can be constructed as a quotient of part of the original structure via an equivalence relation. An important example is a quotient group of a group.

One might say that to understand the full structure one must understand these quotients. When the equivalence relation is definable, we can give the previous sentence a precise meaning. We say that these structures are interpretable.

A key fact is that one can translate sentences from the language of the interpreted structures to the language of the original structure. Thus one can show that if a structure M interprets another whose theory is undecidable, then M itself is undecidable.

Using the Compactness and Completeness Theorems

Gödel's completeness theorem says that a theory has a model if and only if it is consistent, i.e. no contradiction is proved by the theory. This is the heart of model theory as it lets us answer questions about theories by looking at models and vice versa. One should not confuse the completeness theorem with the notion of a complete theory. A complete theory is a theory that contains every sentence or its negation. Importantly, one can find a complete consistent theory extending any consistent theory. However, as shown by Gödel's incompleteness theorems only in relatively simple cases will it be possible to have a complete consistent theory that is also recursive, i.e. that can be described by a recursively enumerable set of axioms. In particular, the theory of natural numbers has no recursive complete and consistent theory. Non-recursive theories are of little practical use, since it is undecidable if a proposed axiom is indeed an axiom, making proof-checking a supertask.

The compactness theorem states that a set of sentences S is satisfiable if every finite subset of S is satisfiable. In the context of proof theory the analogous statement is trivial, since every proof can have only a finite number of antecedents used in the proof. In the context of model theory, however, this proof is somewhat more difficult. There are two well known proofs, one by Gödel (which goes via proofs) and one by Malcev (which is more direct and allows us to restrict the cardinality of the resulting model).

Model theory is usually concerned with first-order logic, and many important results (such as the completeness and compactness theorems) fail in second-order logic or other alternatives. In first-order logic all infinite cardinals look the same to a language which is countable. This is expressed in the Löwenheim–Skolem theorems, which state that any countable theory with an infinite model $\mathfrak{A}$ has models of all infinite cardinalities (at least that of the language) which agree with $\mathfrak{A}$ on all sentences, i.e. they are 'elementarily equivalent'.

Types

Fix an L-structure M, and a natural number n. The set of definable subsets of M^n over some parameters A is a Boolean algebra. By Stone's representation theorem for Boolean algebras there is a natural dual notion to this. One can consider this to be the topological space consisting of maximal consistent sets of formulae over A. We call this the space of (complete) n-types over A, and write $S_n(A)$.

Now consider an element $m \in M^n$. Then the set of all formulae ϕ with parameters in A in free variables $x_1,...,x_n$ so that $M \vDash \phi(m)$ is consistent and maximal such. It is called the *type* of m over A.

One can show that for any n-type p, there exists some elementary extension N of M and some $a \in N^n$ so that p is the type of a over A.

Many important properties in model theory can be expressed with types. Further many proofs go via constructing models with elements that contain elements with certain types and then using these elements.

Illustrative Example: Suppose M is an algebraically closed field. The theory has quantifier elimination. This allows us to show that a type is determined exactly by the polynomial equations it contains. Thus the space of n-types over a subfield A is bijective with the set of prime ideals of the polynomial ring $A[x_1,...,x_n]$. This is the same set as the spectrum of $A[x_1,...,x_n]$. Note however that the topology considered on the type space is the constructible topology: a set of types is basic open iff it is of the form $\{p : f(x) = 0 \in p\}$ or of the form $\{p : f(x) \neq 0 \in p\}$. This is finer than the Zariski topology.

Recursion Theory

Recursion theory deals with the fundamental concepts on what subsets of natural numbers (or other famous countable domains) could be defined effectively and how complex the so defined sets are. The basic concept are the recursive and recursively enumerable sets, but the world of sets investigated in recursion theory goes beyond these sets. The notions are linked to Diophantine sets, definability by functions via recursion and turing machines.

For a long time the term "recursion" was used by mathematicians without being accurately defined. Its approximate intuitive sense can be described in the following way: The value of a sought function f at an arbitrary point $\bar{x}$ (by point is understood a tuple of values of arguments) is determined, generally speaking, by way of the values of this same function at other points $\bar{y}$ that in a sense "precede" $\bar{x}$. (The very word "recur" means "return".) At certain "initial" points the values of f must of course be defined directly. Sometimes recursion is used to define several functions simultaneously; then the above-mentioned definitions are taken with a corresponding modification. Examples of different kinds of recursion will be given below. The relation "x1 precedes x2" (where $\bar{x}_1, \bar{x}_2$ belong to the domain of the sought function) in various types of recursion ("recursive schemes") may have a different sense. It must, however, be "well-founded recursionwell-founded" (i.e. there should not be an infinite sequence of points $\bar{x}_n, n = 0,1 ,...,$ such that $\bar{x}_n, +1$ precedes $\bar{x}_n,$). Furthermore, it is implicitly understood that the relation is "sufficiently natural" (e.g. it is desirable that this relation be seen from the actual description of a recursive scheme and not from the process of its application). This last condition has a purely heuristic value (e.g. for determining any special, comparatively simple type of recursion). A more accurate definition of this condition is essentially inseparable from a more accurate description of the concept of recursion itself, and for this it is essential to establish which type of formal expressions can be acknowledged as recursive definitions.

Where it is a question of recursive descriptions of numerical functions (i.e. functions with natural arguments and natural values), it is usually implied that such descriptions define a way of computing the functions being determined. Here and in the sequel (except in the concluding remarks), the term "recursion" will be understood in precisely this sense. The simplest and most widely used recursive scheme is primitive recursion:

$$f(0, x1 ,..., x_n) = g(x1,..., x_n),$$

$$f(y+1, x1 ,..., x_n)$$

$$= h(y, f(y, x1 ,..., x_n), x1 ,..., x_n),$$

Where the functions g and h are assumed to be known, f is the function to be determined, y is a variable according to which the recursion is conducted, and $x1 ,..., x_n$ are parameters not participating in the recursion. The closest generalization of this scheme is the so-called course-of-value recursion, which includes those types of recursive definitions in which, as in primitive recursion, only one variable participates in the recursion. The corresponding relation of precedence coincides with the normal ordering of the natural numbers (sometimes, however, this term is used in an even wider sense). The most typical form of course-of-value recursion is as follows:

$$f(0, x1 ,..., x_n) = g(x1 ,..., x_n)$$

$$f(y+1, x1 ,..., x_n)$$

$$= h(y, f(\alpha1 (y), x1 ,..., x_n),...,$$

$$,\ldots,\ f\big(\alpha\kappa\,(y),x1\ ,\ldots,\ x_n\big),x1\ ,\ldots,\ x_n\big),$$

Where, $\alpha i\,(y)\le y, i=1\ ,..,k$. Mathematical logic often involves primitive recursive functions, i.e. functions that can be obtained after a finite number of steps using substitution and primitive recursion, starting from a specific fixed supply of basic functions (e.g. $f\,(x)=x+1, f\,(x,y)=y$, etc.). A sequence of functional equalities that describes such a structure is called a primitive recursive description of the corresponding function. These descriptions are syntactic objects (i.e. chains of symbols) possessing a definite effectively recognizable structure. Practically all numerical functions used in mathematics for some concrete purpose prove to be primitive recursive. This largely explains the interest there is in this class of functions.

More complex types of recursive definitions are obtained when the recursion occurs simultaneously over several variables. These definitions, as a rule, lead out of the class of primitive recursive functions, although the corresponding relation of precedence may be completely natural. For example, the values of $f(u,v)$ may participate in the definition of $f(x,y)$, where $u<x$ or $u=x$, $v<y$, as this occurs in the following scheme of double recursion:

$$f(x,y)=g(x,y)\quad\text{if } x=0 \text{ or } y=0,$$

$$f(x+1,y+1)$$

$$=h\Big(x,y,f,\big(x,\alpha\,(x,y,f(x+1,y))\big)\Big),\ f\big(x+1,y\big)\Big).$$

This scheme has not yet been reduced to primitive recursion. On the other hand, a lot of more involved recursive definitions have been reduced to it.

The partial types of recursion referred to have precise mathematical definitions, as opposed to the vague "near mathematical" ideas about "recursion in general" . The refinement of these ideas is naturally thought of as a search for a suitable algorithmic language (i.e. a formal language for describing computable procedures) that incorporates all the imaginable types of recursion without being too broad. One can rightly expect that the development of such a refinement may require further additional agreements, which bring something new to the intuitive understanding of recursion. In this connection it is not without interest to note that all the recursive schemes examined above are oriented towards the generation of total (everywhere defined) functions. Thus, if in a scheme of primitive recursion the functions g, h are total, then so is f. In general, recursive definitions that define partial functions intuitively look rather unnatural. However, such recursions are brought into the analysis for a definite reason, connected with the so-called diagonal method. Thus, let there be given an algorithmic language L in which only total functions are defined, and let the syntactic structure of L not extend beyond the limits of an intuitively understood recursiveness. It is of course implied that expressions in the language L (being descriptions of functions) are algorithmically recognizable and that there is a uniform method of computing functions that can be represented in L, according to their descriptions. If expressions in L are regarded as being simply chains of symbols, then each of these chains can be considered as a representation of a natural number in a suitable number system, the "code" of the given expression. Now, let the function $G(m,x)$ be defined as follows: If m is the code for describing (in L) a certain one-place function

fm, then $G(m,x) = fm(x)$. Otherwise $G(m,x) = \phi(x),$, where ϕ is some fixed function representable in L. It is clear that $G(m,x)$ is computable and total, as is the function $F(x) = G(x,x) + 1$. But F is not expressible in L, since for any m the equality $F(m) = fm(m)$ is impossible. (This argument makes considerable use of the totality of F.) The question arises: Is the given description of F recursive? If L is taken to be the language of primitive recursive descriptions, then the description turns out to be reducible to a double recursion. Generally, the situation is unclear because of the vagueness in intuitive ideas about recursion, so that a positive answer to this question appears to be an additional agreement. If it is accepted (as is implicitly done in contemporary logic), then the existence of a language L that describes in detail all types of "total" recursion proves to be impossible.

Meanwhile, if partial functions can also be expressed in L (and if their descriptions are acknowledged to be recursive), then diagonalization does not necessarily extend beyond L, so that this language may be suitable for an adequate precization of recursiveness. It is true that this involves a certain re-thinking of the very concept of recursion, and the contemporary strict definition of this concept is not fully in line with previously held intuitive ideas. In the new refined sense, recursion is a specific syntactic operation (with a fixed interpretation) used to construct expressions in various algorithmic languages. If L is one of these languages, it is natural to assume that there are other syntactic operations in it on which the concrete type of recursive descriptions in L depends. Generally speaking, expressions in L do not necessarily describe only numerical functions. Several of them may define functional operators and other objects. This is necessary, in particular, for defining recursion. Recursive descriptions in L are systems of functional equations of the type $f_i = T_i(f_1, ..., f_n), i-1, ...n$, where f are the functions being defined and the T_i are expressions in L defining a specific operator in a collection. But all expressible operators in L must be effective (since L is an algorithmic language) and hence monotone (i.e. they preserve the relation $\phi \prec \psi$,, which means that ψ supplements the definition of ϕ). Because of this, every system of the given type has a minimal (in the sense of the relation $\prec$) solution and, according to the definition, is a recursive description of the functions that make up this solution. Starting from the given description, the sought minimal solution can be obtained by means of the following process. For $i = 1, ..., n$, let

f_i^o be a nowhere-defined function and let $f_i^{k+1} = T_i\left(f_1^k, ..., f_n^k\right)$ be such that $f_i^k \prec f_i^{k+1}$. It is easy to

verify that the sought functions f are obtained by "combination" of the $f_i^k (k = 0, 1, ...)$. Methods for a simultaneous computation of f are defined in the same way. This process defines a more or less natural relation of precedence on arguments, which also justifies (to a satisfactory extent) the use of the term "recursion" in this connection.

In order to include the aforementioned recursive schemes in this definition, it is essential that the language L be not too poor. Thus, already in the case of primitive recursion, substitutions and a "piecewise" definition have been required (i.e. a definition of the function by several equalities). Furthermore, these two syntactic operations, in combination with the recursion just defined, are already sufficient to obtain all computable functions (starting from the basic ones). With appropriate assumptions about L, one can confidently claim that the given definition of recursion includes all intuitively conceivable recursive descriptions. At the same time, the general definition given has the characteristic properties of an informally understood recursiveness that are acknowledged to be fundamental in modern mathematics.

The fruitfulness of the given definition of recursion consists not only in its significance in the theory of algorithms, but also in that it permits one to look from an "algorithmic" (in the generalized sense) point of view at several structures of abstract mathematics that have a definite similarity with "numerical" recursions (transfinite recursion, inductive definition, recursive hierarchies, etc.).

References

- Hodges, Wilfrid (1997). A shorter model theory. Cambridge: Cambridge University Press. ISBN 978-0-521-58713-6

- Mathematical-logic, entry: newworldencyclopedia.org, Retrieved 19 April 2018

- Marker, David (2002). Model Theory: An Introduction. Graduate Texts in Mathematics 217. Springer. ISBN 0-387-98760-6

- Set-theory, science: britannica.com, Retrieved 29 June 2018

- Hazewinkel, Michiel, ed. (2001) [1994], "Model theory", Encyclopedia of Mathematics, Springer Science+Business Media B.V. / Kluwer Academic Publishers, ISBN 978-1-55608-010-4

- Constructive, mathematics: ncatlab.org, Retrieved 30 June 2018

- Ebbinghaus, Heinz-Dieter; Flum, Jörg; Thomas, Wolfgang (1994). Mathematical Logic. Springer. ISBN 0-387-94258-0

- Model-theory, mathematical-logic, research-groups: maths.manchester.ac.uk, Retrieved 03 June 2018

- Rautenberg, Wolfgang (2010). A Concise Introduction to Mathematical Logic (3rd ed.). New York: Springer Science+Business Media. doi:10.1007/978-1-4419-1221-3. ISBN 978-1-4419-1220-6

Theorems in Mathematical Logic

An understanding of the field of mathematical logic requires an analysis of related theories. These include Gödel's completeness and incompleteness theorem, compactness theorem, Morley's categoricity theorem and Lindström's theorem, among others. This chapter closely examines these important theories for a comprehensive understanding of mathematical logic.

Gödel's Completeness Theorem

The following statement on the completeness of classical predicate calculus: Any predicate formula that is true in all models is deducible (by formal rules of classical predicate calculus). The theorem proves that the set of deducible formulas of this calculus is, in a certain sense, maximal: it contains all purely-logical laws of set-theoretic mathematics. K. Gödel's proof yields a means of constructing a countermodel (i.e. model for the negation) of any formula A that is not-deducible in the Gentzen formal system without cut-rule. There are also proofs based on extensions of the system of formulas to a maximal system, and also proofs involving the use of algebraic methods. The theorem together with the proof has been generalized to include calculi with equality. Another direction is a generalization to arbitrary sets of formulas: Any consistent set of formulas has a model (a set M is called consistent if for any $A_1, ..., A_k \subset M$ it is not possible to deduce $\neg A_1 \wedge ... \wedge A_k$). Gödel's proof yields a model, with terms as its elements, for a consistent set of formulas. Such models constitute a starting point in many meta-mathematical studies on set theory. Another application of models of terms is the Löwenheim–Skolem theorem: If a denumerable set of formulas has a model, then it has a denumerable model. Gödel's proof itself can be carried out in set theory without the axiom of infinity, i.e. by arithmetical means. One obtains thus a constructive form of Gödel's completeness theorem (Bernays' lemma): For each predicate formula A it is possible to find a substitution ξ of arithmetical predicates for predicate variables such that $\xi A \to \mathrm{Pr}(A)$ is deducible in formal arithmetic. Here $\mathrm{Pr}(A)$ is the arithmetical formula that says that A is deducible. Thus, for A to be deducible it is sufficient for it to be true in the model defined by the substitution ξ. Bernays' lemma is used to construct models of a formal system S in a system S' if in S' the consistency of S has been proved.

From Gödel's completeness theorem one may deduce the cut-elimination theorem and various separation theorems, e.g.: If a formula not containing the equality sign is deducible in predicate calculus with equality, then it is also deducible by pure predicate calculus; if a predicate formula is deducible in arithmetic with free predicate variables, then it is deducible in predicate calculus.

Gödel's completeness theorem may be generalized (if the concept of a model is suitably generalized as well) to non-classical calculi: intuitionistic, modal, etc., as well as to infinitary languages.

Theorem 1 (Gödel completeness theorem, informal statement) Let Γ be a first-order theory (a formal language $\mathcal{L}$, together with a set of axioms, i.e. sentences assumed to be true), and let ϕ be a sentence in the formal language. Assume also that the language $\mathcal{L}$ has at most countably many symbols. Then the following are equivalent:

- (Syntactic consequence) ϕ can be deduced from the axioms in Γ by a finite number of applications of the laws of deduction in first order logic. (This property is abbreviated as $\Gamma \mid- \phi$.)

- (Semantic consequence) Every structure $\mathfrak{U}$ which satisfies or *models* Γ, also satisfies ϕ. (This property is abbreviated as $\Gamma \models \phi$.)

- (Semantic consequence for at most countable models) Every structure $\mathfrak{U}$ which is at most countable, and which models Γ, also satisfies ϕ.

Gödel's Original Formulation

The completeness theorem says that if a formula is logically valid then there is a finite deduction (a formal proof) of the formula.

Gödel's completeness theorem says that a deductive system of first-order predicate calculus is "complete" in the sense that no additional inference rules are required to prove all the logically valid formulas. A converse to completeness is soundness, the fact that only logically valid formulas are provable in the deductive system. Together with soundness (whose verification is easy), this theorem implies that a formula is logically valid if and only if it is the conclusion of a formal deduction.

Model Existence Theorem

The simplest version of this theorem that suffices in practice for most needs, and has connections with the Löwenheim–Skolem theorem, says:

> Every consistent, countable first-order theory has a finite or countable model.

A more general version can be expressed as:

> Every consistent first-order theory with a well-orderable language has a model.

Here, a consistent theory is defined as one in which, for no formula F, both F and $\neg F$ can be proven.

This theorem by Henkin is the most directly obtained version of the completeness theorem in its simplest proof.

Given Henkin's theorem, the proof of the completeness theorem is as follows: If $\Phi \models A$ is valid, then $\Phi \cup \neg A$ does not have models. By the contrapositive of Henkin's, then $\neg A$ is an inconsistent formula. But, by the definition of consistency, if $\Phi \cup \neg A$ is inconsistent then it's possible to build a proof of $\Phi \vdash A$.

More General Form

It says that for any first-order theory T with a well-orderable language, and any sentence s in the language of the theory, there is a formal proof of s in T if and only if s is satisfied by every model of T (s is a semantic consequence of T).

This more general theorem is used implicitly, for example, when a sentence is shown to be provable from the axioms of group theory by considering an arbitrary group and showing that the sentence is satisfied by that group. It is deduced from the model existence theorem as follows: if there is no formal proof of a formula then adding its negation to the axioms gives a consistent theory, which has thus a model, so that the formula is not a semantic consequence of the initial theory.

Gödel's original formulation is deduced by taking the particular case of a theory without any axiom.

As a Theorem of Arithmetic

The Model Existence Theorem and its proof can be formalized in the framework of Peano arithmetic. Precisely, we can systematically define a model of any consistent effective first-order theory T in Peano arithmetic by interpreting each symbol of T by an arithmetical formula whose free variables are the arguments of the symbol. However, the definition expressed by this formula is not recursive.

Consequences

An important consequence of the completeness theorem is that it is possible to recursively enumerate the semantic consequences of any effective first-order theory, by enumerating all the possible formal deductions from the axioms of the theory, and use this to produce an enumeration of their conclusions.

This comes in contrast with the direct meaning of the notion of semantic consequence, that quantifies over all structures in a particular language, which is clearly not a recursive definition.

Also, it makes the concept of "provability," and thus of "theorem," a clear concept that only depends on the chosen system of axioms of the theory, and not on the choice of a proof system.

Relationship to the Incompleteness Theorem

Gödel's incompleteness theorem, another celebrated result, shows that there are inherent limitations in what can be achieved with formal proofs in mathematics. The name for the incompleteness theorem refers to another meaning of *complete*.

It shows that in any consistent effective theory T containing Peano arithmetic (PA), the formula C_T expressing the consistency of T cannot be proven within T.

Applying the completeness theorem to this result, gives the existence of a model of T where the formula C_T is false. Such a model (precisely, the set of "natural numbers" it contains) is necessarily non-standard, as it contains the code number of a proof of a contradiction of T. But T is consistent when viewed from the outside. Thus this code number of a proof of contradiction of T must be a non-standard number.

In fact, the model of *any* theory containing PA obtained by the systematic construction of the arithmetical model existence theorem, is *always* non-standard with a non-equivalent provability predicate and a non-equivalent way to interpret its own construction, so that this construction is non-recursive (as recursive definitions would be unambiguous).

Also, there is no recursive non-standard model of PA.

Relationship to the Compactness Theorem

The completeness theorem and the compactness theorem are two cornerstones of first-order logic. While neither of these theorems can be proven in a completely effective manner, each one can be effectively obtained from the other.

The compactness theorem says that if a formula φ is a logical consequence of a (possibly infinite) set of formulas Γ then it is a logical consequence of a finite subset of Γ. This is an immediate consequence of the completeness theorem, because only a finite number of axioms from Γ can be mentioned in a formal deduction of φ, and the soundness of the deduction system then implies φ is a logical consequence of this finite set. This proof of the compactness theorem is originally due to Gödel.

Conversely, for many deductive systems, it is possible to prove the completeness theorem as an effective consequence of the compactness theorem.

The ineffectiveness of the completeness theorem can be measured along the lines of reverse mathematics. When considered over a countable language, the completeness and compactness theorems are equivalent to each other and equivalent to a weak form of choice known as weak König's lemma, with the equivalence provable in RCA_0 (a second-order variant of Peano arithmetic restricted to induction over Σ_1^0 formulas). Weak König's lemma is provable in ZF, the system of Zermelo–Fraenkel set theory without axiom of choice, and thus the completeness and compactness theorems for countable languages are provable in ZF. However the situation is different when the language is of arbitrary large cardinality since then, though the completeness and compactness theorems remain provably equivalent to each other in ZF, they are also provably equivalent to a weak form of the axiom of choice known as the ultrafilter lemma. In particular, no theory extending ZF can prove either the completeness or compactness theorems over arbitrary (possibly uncountable) languages without also proving the ultrafilter lemma on a set of same cardinality, knowing that on countable sets, the ultrafilter lemma becomes equivalent to weak König's lemma.

Gödel's Incompleteness Theorem

Gödel's two incompleteness theorems are among the most important results in modern logic, and have deep implications for various issues. They concern the limits of provability in formal axiomatic theories. The first incompleteness theorem states that in any consistent formal system F within which a certain amount of arithmetic can be carried out, there are statements of the language of F which can neither be proved nor disproved in F. According to the second incompleteness theorem, such a formal system cannot prove that the system itself is consistent (assuming it is indeed consistent). These results have had a great impact on the philosophy of mathematics and logic. There have been attempts to apply the results also in other areas of philosophy such as the philosophy of mind, but these attempted applications are more controversial. The present entry surveys the two incompleteness theorems and various issues surrounding them.

Gödel's incompleteness theorems are among the most important results in modern logic. These discoveries revolutionized the understanding of mathematics and logic, and had dramatic implications for the philosophy of mathematics. There have also been attempts to apply them in other fields of philosophy, but the legitimacy of many such applications is much more controversial.

In order to understand Gödel's theorems, one must first explain the key concepts essential to it, such as "formal system", "consistency", and "completeness". Roughly, a *formal system* is a system of axioms equipped with rules of inference, which allow one to generate new theorems. The set of axioms is required to be finite or at least decidable, i.e., there must be an algorithm (an effective method) which enables one to mechanically decide whether a given statement is an axiom or not. If this condition is satisfied, the theory is called "recursively axiomatizable", or, simply, "axiomatizable". The rules of inference (of a formal system) are also effective operations, such that it can always be mechanically decided whether one has a legitimate application of a rule of inference at hand. Consequently, it is also possible to decide for any given finite sequence of formulas, whether it constitutes a genuine derivation, or a proof, in the system—given the axioms and the rules of inference of the system.

A formal system is *complete* if for every statement of the language of the system, either the statement or its negation can be derived (i.e., proved) in the system. A formal system is *consistent* if there is no statement such that the statement itself and its negation are both derivable in the system. Only consistent systems are of any interest in this context, for it is an elementary fact of logic that in an inconsistent formal system every statement is derivable, and consequently, such a system is trivially complete.

Gödel established two different though related incompleteness theorems, usually called the first incompleteness theorem and the second incompleteness theorem. "Gödel's theorem" is sometimes used to refer to the conjunction of these two, but may refer to either—usually the first—separately. Accommodating an improvement due to J. Barkley Rosser in 1936, the first theorem can be stated, roughly, as follows:

First Incompleteness Theorem

Any consistent formal system F within which a certain amount of elementary arithmetic can be carried out is incomplete; i.e., there are statements of the language of F which can neither be proved nor disproved in F.

Gödel's theorem does not merely claim that such statements exist: the method of Gödel's proof explicitly produces a particular sentence that is neither provable nor refutable in F; the "undecidable" statement can be found mechanically from a specification of F. The sentence in question is a relatively simple statement of number theory, a purely universal arithmetical sentence.

A common misunderstanding is to interpret Gödel's first theorem as showing that there are truths that cannot be proved. This is, however, incorrect, for the incompleteness theorem does not deal with provability in any absolute sense, but only concerns derivability in some particular formal system or another. For any statement A unprovable in a particular formal system F, there are, trivially, other formal systems in which A is provable (take A as an axiom). On the other hand, there is the extremely powerful standard axiom system of Zermelo-Fraenkel set theory, which is more than sufficient for the derivation of all ordinary mathematics. Now there are, by Gödel's first theorem, arithmetical truths that are not provable even in ZFC. Proving them would thus require a formal system that incorporates methods going beyond ZFC. There is thus a sense in which such truths are not provable using today's "ordinary" mathematical methods and axioms, nor can they be proved in a way that mathematicians would today regard as unproblematic and conclusive.

Syntactic Form of the Gödel Sentence

The Gödel sentence is designed to refer, indirectly, to itself. The sentence states that, when a particular sequence of steps is used to construct another sentence, that constructed sentence will not be provable in F. However, the sequence of steps is such that the constructed sentence turns out to be G_F itself. In this way, the Gödel sentence G_F indirectly states its own unprovability within F

To prove the first incompleteness theorem, Gödel demonstrated that the notion of provability within a system could be expressed purely in terms of arithmetical functions that operate on Gödel numbers of sentences of the system. Therefore, the system, which can prove certain facts about numbers, can also indirectly prove facts about its own statements, provided that it is effectively generated. Questions about the provability of statements within the system are represented as questions about the arithmetical properties of numbers themselves, which would be decidable by the system if it were complete.

Thus, although the Gödel sentence refers indirectly to sentences of the system F, when read as an arithmetical statement the Gödel sentence directly refers only to natural numbers. It asserts that no natural number has a particular property, where that property is given by a primitive recursive relation. As such, the Gödel sentence can be written in the language of arithmetic with a simple syntactic form. In particular, it can be expressed as a formula in the language of arithmetic consisting of a number of leading universal quantifiers followed by a quantifier-free body (these formulas are at level $\prod_1^0$ of the arithmetical hierarchy). Via the MRDP theorem, the Gödel sentence can be re-written as a statement that a particular polynomial in many variables with integer coefficients never takes the value zero when integers are substituted for its variables.

Truth of the Gödel Sentence

The first incompleteness theorem shows that the Gödel sentence G_F of an appropriate formal theory F is unprovable in F. Because, when interpreted as a statement about arithmetic, this unprovability is exactly what the sentence (indirectly) asserts, the Gödel sentence is, in fact, true. For this reason, the sentence G_F is often said to be "true but unprovable." (Raatikainen 2015). However, since the Gödel sentence cannot itself formally specify its intended interpretation, the truth of the sentence G_F may only be arrived at via a meta-analysis from outside the system. In general, this meta-analysis can be carried out within the weak formal system known as primitive recursive arithmetic, which proves the implication $\text{Con}(F) \to G_F$, where $\text{Con}(F)$ is a canonical sentence asserting the consistency of F.

Although the Gödel sentence of a consistent theory is true as a statement about the intended interpretation of arithmetic, the Gödel sentence will be false in some nonstandard models of arithmetic, as a consequence of Gödel's completeness theorem. That theorem shows that, when a sentence is independent of a theory, the theory will have models in which the sentence is true and models in which the sentence is false. As described earlier, the Gödel sentence of a system F is an arithmetical statement which claims that no number exists with a particular property. The incompleteness theorem shows that this claim will be independent of the system F, and the truth of the Gödel sentence follows from the fact that no standard natural number has the property in question. Any model in which the Gödel sentence is false must contain some element which satisfies the property within that model. Such a model must be "nonstandard" – it must contain elements that do not correspond to any standard natural number.

Relationship with the Liar Paradox

Gödel specifically cites Richard's paradox and the liar paradox as semantical analogues to his syntactical incompleteness result in the introductory section of "On Formally Undecidable Propositions in Principia Mathematica and Related Systems I". The liar paradox is the sentence "This sentence is false." An analysis of the liar sentence shows that it cannot be true (for then, as it asserts, it is false), nor can it be false (for then, it is true). A Gödel sentence G for a system F makes a similar assertion to the liar sentence, but with truth replaced by provability: G says "G is not provable in the system F." The analysis of the truth and provability of G is a formalized version of the analysis of the truth of the liar sentence.

It is not possible to replace "not provable" with "false" in a Gödel sentence because the predicate "Q is the Gödel number of a false formula" cannot be represented as a formula of arithmetic. This result, known as Tarski's undefinability theorem, was discovered independently both by Gödel, when he was working on the proof of the incompleteness theorem, and by the theorem's namesake, Alfred Tarski.

Extensions of Gödel's Original Result

Compared to the theorems stated in Gödel's 1931 paper, many contemporary statements of the incompleteness theorems are more general in two ways. These generalized statements are phrased to apply to a broader class of systems, and they are phrased to incorporate weaker consistency assumptions.

Gödel demonstrated the incompleteness of the system of *Principia Mathematica*, a particular system of arithmetic, but a parallel demonstration could be given for any effective system of a certain expressiveness. Gödel commented on this fact in the introduction to his paper, but restricted the proof to one system for concreteness. In modern statements of the theorem, it is common to state the effectiveness and expressiveness conditions as hypotheses for the incompleteness theorem, so that it is not limited to any particular formal system. The terminology used to state these conditions was not yet developed in 1931 when Gödel published his results.

Gödel's original statement and proof of the incompleteness theorem requires the assumption that the system is not just consistent but *ω-consistent*. A system is ω-consistent if it is not ω-inconsistent, and is ω-inconsistent if there is a predicate P such that for every specific natural number m the system proves $\sim P(m)$, and yet the system also proves that there exists a natural number n such that $P(n)$. That is, the system says that a number with property P exists while denying that it has any specific value. The ω-consistency of a system implies its consistency, but consistency does not imply ω-consistency. J. Barkley Rosser (1936) strengthened the incompleteness theorem by finding a variation of the proof (Rosser's trick) that only requires the system to be consistent, rather than ω-consistent. This is mostly of technical interest, because all true formal theories of arithmetic (theories whose axioms are all true statements about natural numbers) are ω-consistent, and thus Gödel's theorem as originally stated applies to them. The stronger version of the incompleteness theorem that only assumes consistency, rather than ω-consistency, is now commonly known as Gödel's incompleteness theorem and as the Gödel–Rosser theorem.

Gödel's second incompleteness theorem concerns the limits of consistency proofs. A rough statement is:

Second Incompleteness Theorem

For any consistent system F within which a certain amount of elementary arithmetic can be carried out, the consistency of F cannot be proved in F itself.

In the case of the second theorem, F must contain a little bit more arithmetic than in the case of the first theorem, which holds under very weak conditions. It is important to note that this result, like the first incompleteness theorem, is a theorem about formal provability, or derivability (which is always relative to some formal system; in this case, to F itself). It does not say anything about whether, for a particular theory T satisfying the conditions of the theorem, the statement "T is consistent" can be proved in the sense of being shown to be true by a conclusive argument, or by a proof generally acceptable for mathematicians

Expressing Consistency

There is a technical subtlety in the second incompleteness theorem regarding the method of expressing the consistency of F as a formula in the language of F. There are many ways to express the consistency of a system, and not all of them lead to the same result. The formula $\text{Cons}(F)$ from the second incompleteness theorem is a particular expression of consistency.

Other formalizations of the claim that F is consistent may be inequivalent in F, and some may even be provable. For example, first-order Peano arithmetic (PA) can prove that "the largest consistent subset of PA" is consistent. But, because PA is consistent, the largest consistent subset of PA is just PA, so in this sense PA "proves that it is consistent". What PA does not prove is that the largest consistent subset of PA is, in fact, the whole of PA. (The term "largest consistent subset of PA" is meant here to be the largest consistent initial segment of the axioms of PA under some particular effective enumeration).

The Hilbert–Bernays Conditions

The standard proof of the second incompleteness theorem assumes that the provability predicate $\text{Prov}_A(P)$ satisfies the Hilbert–Bernays provability conditions. Letting $\#(P)$ represent the Gödel number of a formula P, the derivability conditions say:

1. If F proves P, then F proves $Prov_A\big(\#(P)\big)$.

2. F proves 1.; that is, F proves that if F proves P, then F proves $Prov_A(\#(P))$. In other words, F proves that $Prov_A(\#(P))$ *implies* $Prov_A(\#(Prov_A\big(\#(P)\big)))$.

3. F proves that if F proves that $(P \rightarrow Q)$ and F proves P then F proves Q. In other words, F proves that $Prov_A(\#(P \rightarrow Q))$ *and* $Prov_A(\#(P))$ *imply* $Prov_A(\#(Q))$.

There are systems, such as Robinson arithmetic, which are strong enough to meet the assumptions of the first incompleteness theorem, but which do not prove the Hilbert—Bernays conditions. Peano arithmetic, however, is strong enough to verify these conditions, as are all theories stronger than Peano arithmetic.

Implications for Consistency Proofs

Gödel's second incompleteness theorem also implies that a system F_1 satisfying the technical conditions outlined above cannot prove the consistency of any system F_2 that proves the consistency of F_1. This is because such a system F_1 can prove that if F_2 proves the consistency of F_1, then F_1 is in fact consistent. For the claim that F_1 is consistent has form "for all numbers n, n has the decidable property of not being a code for a proof of contradiction in F_1". If F_1 were in fact inconsistent, then F_2 would prove for some n that n is the code of a contradiction in F_1. But if F_2 also proved that F_1 is consistent (that is, that there is no such n), then it would itself be inconsistent. This reasoning can be formalized in F_1 to show that if F_2 is consistent, then F_1 is consistent. Since, by second incompleteness theorem, F_1 does not prove its consistency, it cannot prove the consistency of F_2 either.

This corollary of the second incompleteness theorem shows that there is no hope of proving, for example, the consistency of Peano arithmetic using any finitistic means that can be formalized in a system the consistency of which is provable in Peano arithmetic (PA). For example, the system of primitive recursive arithmetic (PRA), which is widely accepted as an accurate formalization of finitistic mathematics, is provably consistent in PA. Thus PRA cannot prove the consistency of PA. This fact is generally seen to imply that Hilbert's program, which aimed to justify the use of "ideal" (infinitistic) mathematical principles in the proofs of "real" (finitistic) mathematical statements by giving a finitistic proof that the ideal principles are consistent, cannot be carried out.

The corollary also indicates the epistemological relevance of the second incompleteness theorem. It would actually provide no interesting information if a system F proved its consistency. This is because inconsistent theories prove everything, including their consistency. Thus a consistency proof of F in F would give us no clue as to whether F really is consistent; no doubts about the consistency of F would be resolved by such a consistency proof. The interest in consistency proofs lies in the possibility of proving the consistency of a system F in some system F' that is in some sense less doubtful than F itself, for example weaker than F. For many naturally occurring theories F and F', such as F = Zermelo—Fraenkel set theory and F' = primitive recursive arithmetic, the consistency of F' is provable in F, and thus F' cannot prove the consistency of F by the above corollary of the second incompleteness theorem.

The second incompleteness theorem does not rule out consistency proofs altogether, only consistency proofs that can be formalized in the system that is proved consistent. For example, Gerhard

Gentzen proved the consistency of Peano arithmetic in a different system that includes an axiom asserting that the ordinal called ε_0 is wellfounded.

Examples of Undecidable Statements

There are two distinct senses of the word "undecidable" in mathematics and computer science. The first of these is the proof-theoretic sense used in relation to Gödel's theorems, that of a statement being neither provable nor refutable in a specified deductive system. The second sense, which will not be discussed here, is used in relation to computability theory and applies not to statements but to decision problems, which are countably infinite sets of questions each requiring a yes or no answer. Such a problem is said to be undecidable if there is no computable function that correctly answers every question in the problem set.

Because of the two meanings of the word undecidable, the term independent is sometimes used instead of undecidable for the "neither provable nor refutable" sense.

Undecidability of a statement in a particular deductive system does not, in and of itself, address the question of whether the truth value of the statement is well-defined, or whether it can be determined by other means. Undecidability only implies that the particular deductive system being considered does not prove the truth or falsity of the statement. Whether there exist so-called "absolutely undecidable" statements, whose truth value can never be known or is ill-specified, is a controversial point in the philosophy of mathematics.

The combined work of Gödel and Paul Cohen has given two concrete examples of undecidable statements (in the first sense of the term): The continuum hypothesis can neither be proved nor refuted in ZFC (the standard axiomatization of set theory), and the axiom of choice can neither be proved nor refuted in ZF (which is all the ZFC axioms *except* the axiom of choice). These results do not require the incompleteness theorem. Gödel proved in 1940 that neither of these statements could be disproved in ZF or ZFC set theory. In the 1960s, Cohen proved that neither is provable from ZF, and the continuum hypothesis cannot be proved from ZFC.

In 1973, Saharon Shelah showed that the Whitehead problem in group theory is undecidable, in the first sense of the term, in standard set theory.

Gregory Chaitin produced undecidable statements in algorithmic information theory and proved another incompleteness theorem in that setting. Chaitin's incompleteness theorem states that for any system that can represent enough arithmetic, there is an upper bound c such that no specific number can be proved in that system to have Kolmogorov complexity greater than c. While Gödel's theorem is related to the liar paradox, Chaitin's result is related to Berry's paradox.

Undecidable Statements Provable in Larger Systems

These are natural mathematical equivalents of the Gödel "true but undecidable" sentence. They can be proved in a larger system which is generally accepted as a valid form of reasoning, but are undecidable in a more limited system such as Peano Arithmetic.

In 1977, Paris and Harrington proved that the Paris–Harrington principle, a version of the infinite Ramsey theorem, is undecidable in (first-order) Peano arithmetic, but can be proved in the stronger system of second-order arithmetic. Kirby and Paris later showed that Goodstein's theorem, a statement about sequences of natural numbers somewhat simpler than the Paris–Harrington principle, is also undecidable in Peano arithmetic.

Kruskal's tree theorem, which has applications in computer science, is also undecidable from Peano arithmetic but provable in set theory. In fact Kruskal's tree theorem (or its finite form) is undecidable in a much stronger system codifying the principles acceptable based on a philosophy of mathematics called predicativism. The related but more general graph minor theorem (2003) has consequences for computational complexity theory.

Relationship with Computability

The incompleteness theorem is closely related to several results about undecidable sets in recursion theory.

Stephen Cole Kleene (1943) presented a proof of Gödel's incompleteness theorem using basic results of computability theory. One such result shows that the halting problem is undecidable: there is no computer program that can correctly determine, given any program P as input, whether P eventually halts when run with a particular given input. Kleene showed that the existence of a complete effective system of arithmetic with certain consistency properties would force the halting problem to be decidable, a contradiction. This method of proof has also been presented by Shoenfield); Charlesworth (1980); and Hopcroft and Ullman (1979).

Franzén explains how Matiyasevich's solution to Hilbert's 10th problem can be used to obtain a proof to Gödel's first incompleteness theorem. Matiyasevich proved that there is no algorithm that, given a multivariate polynomial $p(x_1, x_2,...,x_k)$ with integer coefficients, determines whether there is an integer solution to the equation $p = 0$. Because polynomials with integer coefficients, and integers themselves, are directly expressible in the language of arithmetic, if a multivariate integer polynomial equation $p = 0$ does have a solution in the integers then any sufficiently strong system of arithmetic T will prove this. Moreover, if the system T is ω-consistent, then it will never prove that a particular polynomial equation has a solution when in fact there is no solution in the integers. Thus, if T were complete and ω-consistent, it would be possible to determine algorithmically whether a polynomial equation has a solution by merely enumerating proofs of T until either "p has a solution" or "p has no solution" is found, in contradiction to Matiyasevich's theorem. Moreover, for each consistent effectively generated system T, it is possible to effectively generate a multivariate polynomial p over the integers such that the equation $p = 0$ has no solutions over the integers, but the lack of solutions cannot be proved in T.

Smorynski shows how the existence of recursively inseparable sets can be used to prove the first incompleteness theorem. This proof is often extended to show that systems such as Peano arithmetic are essentially undecidable.

Chaitin's incompleteness theorem gives a different method of producing independent sentences, based on Kolmogorov complexity. Like the proof presented by Kleene that was mentioned above, Chaitin's theorem only applies to theories with the additional property that all their axioms are true in the standard model of the natural numbers. Gödel's incompleteness theorem is distinguished by its applicability to consistent theories that nonetheless include statements that are false in the standard model; these theories are known as ω-inconsistent.

Proof Sketch for the First Theorem

The proof by contradiction has three essential parts. To begin, choose a formal system that meets the proposed criteria:

1. Statements in the system can be represented by natural numbers (known as Gödel numbers). The significance of this is that properties of statements—such as their truth and falsehood—will be equivalent to determining whether their Gödel numbers have certain properties, and that properties of the statements can therefore be demonstrated by examining their Gödel numbers. This part culminates in the construction of a formula expressing the idea that *"statement S is provable in the system"* (which can be applied to any statement "S" in the system).

2. In the formal system it is possible to construct a number whose matching statement, when interpreted, is self-referential and essentially says that it (i.e. the statement itself) is unprovable. This is done using a technique called "diagonalization" (so-called because of its origins as Cantor's diagonal argument).

3. Within the formal system this statement permits a demonstration that it is neither provable nor disprovable in the system, and therefore the system cannot in fact be ω-consistent. Hence the original assumption that the proposed system met the criteria is false.

Arithmetization of Syntax

The main problem in fleshing out the proof described above is that it seems at first that to construct a statement p that is equivalent to "p cannot be proved", p would somehow have to contain a reference to p, which could easily give rise to an infinite regress. Gödel's ingenious technique is to show that statements can be matched with numbers (often called the arithmetization of syntax) in such a way that *"proving a statement"* can be replaced with *"testing whether a number has a given property"*. This allows a self-referential formula to be constructed in a way that avoids any infinite regress of definitions. The same technique was later used by Alan Turing in his work on the Entscheidungsproblem.

In simple terms, a method can be devised so that every formula or statement that can be formulated in the system gets a unique number, called its Gödel number, in such a way that it is possible to mechanically convert back and forth between formulas and Gödel numbers. The numbers involved might be very long indeed (in terms of number of digits), but this is not a barrier; all that matters is that such numbers can be constructed. A simple example is the way in which English is stored as a sequence of numbers in computers using ASCII or Unicode:

• The word HELLO is represented by 72-69-76-76-79 using decimal ASCII, i.e. the number 7269767679.

• The logical statement x=y => y=x is represented by 120-061-121-032-061-062-032-121-061-120 using octal ASCII, i.e. the number 120061121032061062032121061120.

In principle, proving a statement true or false can be shown to be equivalent to proving that the number matching the statement does or doesn't have a given property. Because the formal system is strong enough to support reasoning about *numbers in general*, it can support reasoning about *numbers that represent formulae and statements* as well. Crucially, because the system can support reasoning about *properties of numbers*, the results are equivalent to reasoning about *provability of their equivalent statements*.

Construction of a Statement about "Provability"

Having shown that in principle the system can indirectly make statements about provability, by analyzing properties of those numbers representing statements it is now possible to show how to create a statement that actually does this.

A formula $F(x)$ that contains exactly one free variable x is called a *statement form* or *class-sign*. As soon as x is replaced by a specific number, the statement form turns into a *bona fide* statement, and it is then either provable in the system, or not. For certain formulas one can show that for every natural number n, F(n) is true if and only if it can be proved (the precise requirement in the original proof is weaker, but for the proof sketch this will suffice). In particular, this is true for every specific arithmetic operation between a finite number of natural numbers, such as "2×3=6".

Statement forms themselves are not statements and therefore cannot be proved or disproved. But every statement form $F(x)$ can be assigned a Gödel number denoted by $G(F)$. The choice of the free variable used in the form $F(x)$ is not relevant to the assignment of the Gödel number $G(F)$.

The notion of provability itself can also be encoded by Gödel numbers, in the following way: since a proof is a list of statements which obey certain rules, the Gödel number of a proof can be defined. Now, for every statement p, one may ask whether a number x is the Gödel number of its proof. The relation between the Gödel number of p and x, the potential Gödel number of its proof, is an arithmetical relation between two numbers. Therefore, there is a statement form Bew(y) that uses this arithmetical relation to state that a Gödel number of a proof of y exists:

> Bew(y) = ∃ x (y is the Gödel number of a formula and x is the Gödel number of a proof of the formula encoded by y).

The name Bew is short for *beweisbar*, the German word for "provable"; this name was originally used by Gödel to denote the provability formula just described. Note that "Bew(y)" is merely an abbreviation that represents a particular, very long, formula in the original language of T; the string "Bew" itself is not claimed to be part of this language.

An important feature of the formula Bew(y) is that if a statement p is provable in the system then Bew($G(p)$) is also provable. This is because any proof of p would have a corresponding Gödel number, the existence of which causes Bew($G(p)$) to be satisfied.

Diagonalization

The next step in the proof is to obtain a statement which, indirectly, asserts its own unprovability. Although Gödel constructed this statement directly, the existence of at least one such statement follows from the diagonal lemma, which says that for any sufficiently strong formal system and any statement form F there is a statement p such that the system proves:

$$p \leftrightarrow F(\mathbf{G}(p)).$$

By letting F be the negation of Bew(x), we obtain the theorem:

$$p \leftrightarrow \sim Bew(\mathbf{G}(p))$$

and the p defined by this roughly states that its own Gödel number is the Gödel number of an unprovable formula.

The statement p is not literally equal to $\sim Bew(\mathbf{G}(p))$; rather, p states that if a certain calculation is performed, the resulting Gödel number will be that of an unprovable statement. But when this calculation is performed, the resulting Gödel number turns out to be the Gödel number of p itself. This is similar to the following sentence in English:

> "when preceded by itself in quotes, is unprovable.", when preceded by itself in quotes, is unprovable.

This sentence does not directly refer to itself, but when the stated transformation is made the original sentence is obtained as a result, and thus this sentence indirectly asserts its own unprovability. The proof of the diagonal lemma employs a similar method.

Now, assume that the axiomatic system is ω-consistent, and let p be the statement obtained.

If p were provable, then Bew(G(p)) would be provable, as argued above. But p asserts the negation of Bew(G(p)). Thus the system would be inconsistent, proving both a statement and its negation. This contradiction shows that p cannot be provable.

If the negation of p were provable, then Bew(G(p)) would be provable (because p was constructed to be equivalent to the negation of Bew(G(p))). However, for each specific number x, x cannot be the Gödel number of the proof of p, because p is not provable (from the previous paragraph). Thus on one hand the system proves there is a number with a certain property (that it is the Gödel number of the proof of p), but on the other hand, for every specific number x, we can prove that it does not have this property. This is impossible in an ω-consistent system. Thus the negation of p is not provable.

Thus the statement p is undecidable in our axiomatic system: it can neither be proved nor disproved within the system.

In fact, to show that p is not provable only requires the assumption that the system is consistent. The stronger assumption of ω-consistency is required to show that the negation of p is not provable. Thus, if p is constructed for a particular system:

- If the system is ω-consistent, it can prove neither p nor its negation, and so p is undecidable.

- If the system is consistent, it may have the same situation, or it may prove the negation of p. In the later case, we have a statement ("not p") which is false but provable, and the system is not ω-consistent.

If one tries to "add the missing axioms" to avoid the incompleteness of the system, then one has to add either p or "not p" as axioms. But then the definition of "being a Gödel number of a proof" of a statement changes. which means that the formula Bew(x) is now different. Thus when we apply the diagonal lemma to this new Bew, we obtain a new statement p, different from the previous one, which will be undecidable in the new system if it is ω-consistent.

Proof via Berry's Paradox

George Boolos (1989) sketches an alternative proof of the first incompleteness theorem that uses Berry's paradox rather than the liar paradox to construct a true but unprovable formula. A similar proof method was independently discovered by Saul Kripke. Boolos's proof proceeds by constructing, for any computably enumerable set S of true sentences of arithmetic, another sentence which is true but not contained in S. This gives the first incompleteness theorem as a corollary. According to Boolos, this proof is interesting because it provides a "different sort of reason" for the incompleteness of effective, consistent theories of arithmetic.

Computer Verified Proofs

The incompleteness theorems are among a relatively small number of nontrivial theorems that have been transformed into formalized theorems that can be completely verified by proof assistant software. Gödel's original proofs of the incompleteness theorems, like most mathematical proofs, were written in natural language intended for human readers.

Computer-verified proofs of versions of the first incompleteness theorem were announced by Natarajan Shankar in 1986 using Nqthm (Shankar 1994), by Russell O'Connor in 2003 using Coq (O'Connor 2005) and by John Harrison in 2009 using HOL Light (Harrison 2009). A computer-verified proof of both incompleteness theorems was announced by Lawrence Paulson in 2013 using Isabelle (Paulson 2014).

Proof Sketch for the Second Theorem

The main difficulty in proving the second incompleteness theorem is to show that various facts about provability used in the proof of the first incompleteness theorem can be formalized within the system using a formal predicate for provability. Once this is done, the second incompleteness theorem follows by formalizing the entire proof of the first incompleteness theorem within the system itself.

Let p stand for the undecidable sentence constructed above, and assume that the consistency of the system can be proved from within the system itself. The demonstration above shows that if the system is consistent, then p is not provable. The proof of this implication can be formalized within the system, and therefore the statement "p is not provable", or "not $P(p)$" can be proved in the system.

But this last statement is equivalent to p itself (and this equivalence can be proved in the system), so p can be proved in the system. This contradiction shows that the system must be inconsistent.

Compactness Theorem

A set of formulas Φ is satisfiable iff it is finitely satisfiable.

The compactness theorem is often used in its contrapositive form: A set of formulas Φ is unsatisfiable iff there is some finite subset of Φ that is unsatisfiable.

The theorem is true for both first order logic and propositional logic. The proof for first order logic is outside the scope of this course, but we will give the proof for propositional logic. We will only consider the case of a countable number of propositions. The proof for larger cardinalities is essentially the same, but requires some form of the axiom of choice.

Proof of Compactness Theorem for Propositional Logic

We first prove a useful result about finite satisfiability, which allows us to enlarge a set of formulas while preserving their finite satisfiability.

- Lemma 1 (Finite Satisfiability Lemma)

Let Σ be a finitely satisfiable set of formulas and ϕ be a formula. Then $\Sigma_1 \cup \{\varphi\}$ is finitely satisfiable or $\Sigma_2 \cup \{\neg\varphi\}$ is finitely satisfiable.

Proof: We prove the contrapositive statement. Suppose $\Sigma_1 \cup \{\varphi\}$ and $\Sigma_2 \cup \{\neg\varphi\}$ are both not finitely satisfiable. Then there exist finite subsets Σ_1 and Σ_2 of Σ, such that $\Sigma_1 \cup \{\varphi\}$ is not satisfiable and $\Sigma_2 \cup \{\neg\varphi\}$ is not satisfiable. So models $(\Sigma_1) \cap$ models $(\varphi) = \varnothing$, which means models $(\Sigma_1) \subseteq$ models $(\neg\varphi)$. Similarly, we get models $(\Sigma_2) \subseteq$ models (φ). Then models$(\Sigma 1 \cup \Sigma 2) =$ models $(\Sigma_1) \cap$ models $(\Sigma_2) \subseteq$ models $(\neg\varphi) \cap$ models $(\varphi) = \emptyset$. So $\Sigma 1 \cup \Sigma 2$ is not satisfiable. Since $\Sigma_1 \cup \Sigma_2$ is a finite subset of Σ, therefore Σ is not finitely satisfiable.

- Proof of the compactness theorem

Let Σ be a finitely satisfiable set of propositional formulas and let $AP(\Sigma)$ be countable. Then, by the Relevance Lemma, we can assume $Prop = AP(\Sigma) = \{pi : i \in N\}$ to be a countable set of propositions. We define an increasing sequence of sets of formulas as follows:

$\Sigma_0 = \Sigma$ and for all $i \in N$,

$$\Sigma_{i+1} \begin{cases} \Sigma_i \cup \{p_i\} \, is \, finitely \, satisfiable \\ \Sigma_i \cup \{\neg p_i\}, \, otherwise \end{cases}$$

And we define Σ' to be their union: $\Sigma' = \bigcup_{i=0}^{\infty} \Sigma_i$.

Claim: For all i $i \in N, \Sigma_i$ is finitely satisfiable.

Proof: By induction. The base case is $\Sigma_0 = \Sigma$, which is finitely satisfiable by assumption. Assume that Σ_k is finitely satisfiable for $k \in N$. Then, by Lemma ??, one of $\Sigma_k \cup \{p_k\}$ and $\Sigma_k \cup \{\neg p_k\}$ must be finitely satisfiable, so Σ_{k+1} is also finitely satisfiable.

Claim: Σ' is finitely satisfiable.

Proof: Consider a finite subset of X of Σ'. Since X is finite, and $\Sigma_k \subseteq \Sigma_{k+1}$ for all $k \in N$, there exists a finite $j \in N$, such that $X \subseteq \bigcup_{i=0}^{j} \Sigma_i = \Sigma_j$. Since Σ_j is finitely satisfiable, therefore X is satisfiable.

Now, by the construction of Σ', at least one of p_i and $\neg p_i$ is in Σ' for all i ∈ N. However, Σo is finitely satisfiable, so it cannot contain both pi and $\neg p_i$. Therefore, Σ' contains exactly one of pi and $\neg p_i$ for all $i \in N$. Then we can define a truth assignment $\tau : Prop \to \{0, 1\}$ and a function $v : P\,rop \to$ Literals, where Literals $= \{pi, \neg pi : i \in N\}$, as follows:

$$\tau(p) = \begin{cases} 1', & \text{if } p \in \Sigma' \\ 0, & \text{if } \neg p \in \Sigma \end{cases} \qquad v(p) = \begin{cases} p, & \text{if } p \in \Sigma' \\ \neg p, & \text{if } \neg p \in \Sigma'. \end{cases}$$

Claim: $\tau \models \Sigma$.

Proof: Let $\varphi \in \Sigma$. Define $X = \{v(p) : p \in AP(\varphi)\}$. Then, by the definition of τ and v, we have $\tau \models X$, $X \subseteq \Sigma'$, and $AP(X) = AP(\varphi)$. Since $AP(\varphi)$ is finite, $X \cup \{\varphi\}$ is a finite subset of Σ'. Since Σ' is finitely satisfiable, therefore $X \cup \{\varphi\}$ is satisfiable. Then, by the Relevance Lemma, there exists $\tau' \in 2^{P\,rop}$, $\tau'|_{AP}(\varphi)| = \varphi$ and $\tau'|_{AP(\varphi)}| = X$. Also by the Relevance Lemma, we have $\tau|_{AP(\varphi)}| = X$. Since X contains a literal for each proposition in $AP(\varphi)$, therefore τ and τ' must assign the same value to propositions in $AP(\varphi)$, that is, $\tau|_{AP(\varphi)} = \tau'|_{AP(\varphi)}$. Therefore $\tau|_{AP(\varphi)}| = \varphi$, and by the Relevance Lemma, $\tau| = \varphi$. Since φ was an arbitrary formula in Σ, therefore $\tau| = \Sigma$.

We have shown that Σ is satisfiable. This completes the proof of the theorem.

Applications of Compactness

Definability Results in First Order Logic

Compactness can be used to prove results about the definability of class of structures. For example, we showed that $2C \notin EC$ and $EC_\Delta \cap co - EC_\Delta = EC$.

Proving Theorems Outside Logic

Compactness can also be used to prove results in mathematical fields other than logic. For example, in Assignment 6 you are asked to prove the 3-color version of the following theorem using compactness:

Theorem: A graph is k-colorable *iff* every finite subgraph is k-colorable.

This theorem can then be combined with the famous four color theorem to prove an infinite version of the four color theorem.

Theorem: (Four color theorem). Every finite planar graph is 4-colorable.

Theorem: Every infinite planar graph is 4-colorable.

Proof: Let G be an infinite planar graph. Since every subgraph of a planar graph is also planar, by Theorem ??, every finite subgraph of G is 4-colorable. Then, by Theorem ??, G is also 4-colorable.

Some other examples of theorems that can be proved using compactness:

- König's lemma: Every finitely-branching infinite tree has an infinite path.

- Every partial order can be extended to a total order.

Constructing New Models

In first order logic, compactness is frequently used to construct new and interesting models of familiar structures. Here we give a simple example, where starting from a connected graph G, we construct a non-connected graph G' such that G and G' satisfy the same first order properties.

Consider the infinite graph G whose set of nodes is N (the natural numbers) such that for each $n \in N$, there is an undirected edge from n to $n + 1$. Note that G is a connected graph, since there is a finite path from i to j for $i, j \in N$.

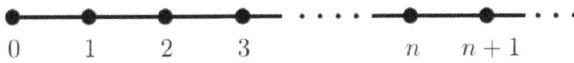

$$G = (V, E) = \left(N, \{(n, n + 1), (n + 1, n) : n \in N\}\right)$$

Let $\Sigma = \{\varphi : G \models \varphi\}$ be the set of all sentences that are true in G. Note that, by definition, $G \models \Sigma$. We define two new constants (0-ary function symbols) a and b, and define $\psi_n = \neg pathn(a, b)$ to be a formula that says there is no path of length n from a to b. Then (V, E, 0, k) satisfies $\Sigma \cup \{\psi_n : n < k\}$, since there is no path of length less than k between node 0 and node k in G (here a is interpreted as 0 and b is interpreted as k). Therefore, $\Sigma' = \Sigma \cup \{\psi_n : n \in N\}$ is finitely satisfiable, and hence, by compactness, Σ' is satisfiable.

Let (V', E', a^0, b^0) be a model of Σ'. Then there is no finite path between a^0 and b^0 in $G' = (V', E')$.. So G' is disconnected. However, because $\Sigma \subseteq \Sigma'$, any model of Σ' is also a model of Σ. Since the symbols a and b do not occur in Σ, we get that G' is a model of Σ. But Σ' was defined to be all sentences satisfied by G. Thus G and G' satisfy the same sentences, that is, $G \equiv G'$. We cannot distinguish between G and G' using first order logic. G' is called a non-standard model of G. Note that G is connected, but G' is not. As a consequence, we immediately get the following results:

Theorem: The class of connected graphs is not EC_Δ.

Proof: Suppose a set of sentences Φ defines the class of connected graphs. Then G is a model of Φ (because it is connected), and therefore $\Phi \subseteq \Sigma$. But then G' is also a model of Φ (because G' is a model of Σ). This contradicts the fact that G' is not connected. Therefore no such set Φ exists and the class of connected graphs is not EC_Δ.

Theorem: Reachability is not definable in first order logic.

Proof: Suppose we could define a formula reach(x, y) to mean that y is reachable from x. Then

$(\forall x)(\forall y)$ reach (x, y) defines the class of connected graphs. But this contradicts the fact that this class is not EC_Δ. Thus, reachability is not definable.

Similarly, non-standard models can be defined for all the familiar mathematical structures such as the natural numbers and the real numbers. For example, one can construct non-standard models of the reals which contain infinitismal numbers that are greater than zero but smaller than all other positive reals.

Löb's Theorem

In 1953, the logician Leon Henkin questioned whether, taking $P(y)$ to be Bew (x) and obtaining $PA \vdash S \leftrightarrow Bew\,(\ulcorner S \urcorner)$ for some sentence S, the sentence S is itself provable in PA. Only a year later, M.H. Löb showed that for all sentences S, if PA 'Bew $(\ulcorner S \urcorner) \to S$, then $PA \vdash S$, a result known as Löb's theorem. This result was not, however, an entirely expected one; there seems to be no a priori reason why PA shouldn't claim to be sound with respect to a proposition which it cannot actually prove; indeed, it even seems natural that $Bew\,(\ulcorner S \urcorner) \to S$ should be true for any S. As Löb's theorem shows, however, PA is incredibly humble in this respect; it never claims to be sound with respect to a proposition unless it must, unless it can actually prove the proposition.

Curry's Paradox

Let SC denote 'Santa Claus exists'. Define $c = \{x : x \in x \to SC\}$. Assume that $c \in c$. Then, by the definition of c, $c \in c \to SC$. Thus, by modus ponens, SC. It follows, then, that $c \in c \to SC$, and so $c \in c$. By modus ponens, then, SC.

A reformulated version of the paradox using Tarski's truth schema:

> 'ϕ' is true if and only if ϕ

is owed to Henkin and often referred to as Henkin's paradox:

Henkin's Paradox

Let SC denote 'Santa Claus exists' and S denote the sentence 'if S is true, then SC'. Assume S is true; then, by definition, 'if S is true, then SC' is true. Thus, by modus ponens, SC. It follows, then, that 'if S is true, then SC' is true–which is to say, S is true. By modus ponens, it follows that SC.

The cyclically self-referential nature of both Curry and Henkin's paradox arises yet again in Löb's theorem.

Theorem

Löb's Theorem

> *If $PA \vdash Bew\,(\ulcorner \varphi \urcorner) \to \phi$, then $PA \vdash \phi$*

Proof:

Applying the diagonal lemma to $\left(Bew(x) \rightarrow \phi\right)$, there is a sentence ψ such that

$$PA \vdash \psi \leftrightarrow (Bew\ (\ulcorner\psi\urcorner) \rightarrow \phi),$$

And so, by definition:

$$PA \vdash \psi \rightarrow (Bew\ (\ulcorner\psi\urcorner) \rightarrow \phi)$$

By property (i) of $Bew(x)$,

$$PA \vdash Bew\ (\ulcorner\psi \rightarrow (Bew\ (\ulcorner\psi\urcorner) \rightarrow \phi)\urcorner)$$

By property (ii) of $Bew(x)$ again,

$$PA \vdash Bew\ (\ulcorner\psi\urcorner) \rightarrow Bew\ (\ulcorner Bew\ (\ulcorner\psi\urcorner) \rightarrow \phi\urcorner)$$

Taking the following instance of property (iii) of $Bew(x)$,

$$PA \vdash Bew\ (\ulcorner\psi\urcorner) \rightarrow Bew\ (\ulcorner Bew\ (\ulcorner\psi\urcorner)\urcorner)$$

Combining the two previous formulas,

$$PA \vdash Bew\ (\ulcorner\psi\urcorner) \rightarrow Bew\ (\ulcorner\phi\urcorner)$$

Assume that $PA \vdash Bew\ (\ulcorner\phi\urcorner) \rightarrow \ulcorner\phi\urcorner$. Then, using the above,

$$PA \vdash Bew\ (\ulcorner\psi\urcorner) \rightarrow \phi$$

But then, by our very first formula,

$$PA \vdash \psi$$

and by property (i) of $Bew(x)$,

$$PA \vdash Bew\ (\ulcorner\psi\urcorner)$$

Finally, by two formulas previous,

$$PA \vdash \phi$$

Corollary (Gödel's Second Incompleteness Theorem)

$$PA \nvdash \neg Bew\ (\ulcorner\bot\urcorner)$$

Proof:

Assume that PA is consistent and that $PA \vdash \neg Bew\left(\ulcorner\bot\urcorner\right)$. Then, trivially, $PA \vdash Bew\left(\ulcorner\bot\urcorner\right) \rightarrow \bot$. But then, by Löb's theorem, $PA \vdash \bot$ — a contradiction. It must be, then, that $PA \nvdash \neg Bew\ (\ulcorner\bot\urcorner)$, and thus the theorem is established.

Löb's theorem in its original form is a generalization of the Gödel incompleteness theorem whose formulation lends itself to the tools of type theory and modal logic.

Löb's theorem states that to prove that a proposition is provable, it is sufficient to prove the proposition under the assumption that it is provable. Since the Curry-Howard isomorphism identifies formal proofs with abstract syntax trees of programs; Löb's theorem implies, for total languages which validate it, that self-interpreters are impossible.

In provability logic the abstract statement is considered in itself as an axiom on a modal operator □\Boxinterpreted as the modality "is provable". In this form the statement reads formally:

$$\Box(\Box P \to P) \to \Box P$$

for any proposition P ("Löb's axiom").

This reduces to an incompleteness theorem when taking P = false and using that

1. negation is $\neg P = (P \to \mathit{false})$;

2. consistency means that $\Box P \to P$

$$\Box(\Box\,\mathit{false} \to \mathit{false}) \to \Box\,\mathit{false}$$
$$\Leftrightarrow\ \Box(\neg\Box\,\mathit{false}) \to \Box\,\mathit{false}$$
$$\Rightarrow\ \Box(\neg\Box\,\mathit{false}) \to \mathit{false}$$
$$\Leftrightarrow\ \neg\Box(\neg\Box\,\mathit{false}$$

Where the last line reads in words "It is not provable that false is not provable."

Modal Proof of Löb's Theorem

Löb's theorem can be proved within modal logic using only some basic rules about the provability operator (the K4 system) plus the existence of modal fixed points.

Modal Formulas

We will assume the following grammar for formulas:

1. If X is a propositional variable, then X is a formula.

2. If K is a propositional constant, then K is a formula.

3. If A is a formula, then $\Box A$ is a formula.

4. If A and B are formulas, then so are $\neg A$, $A \to B$, $A \land B$, $A \lor B$, and $A \leftrightarrow B$.

A modal sentence is a modal formula that contains no propositional variables. We use $\vdash A$ to mean A is a theorem.

Modal Fixed Points

If $F(X)$ is a modal formula with only one propositional variable X, then a modal fixed point of $F(X)$ is a sentence Ψ such that:

$$\vdash \Psi \leftrightarrow F(\Box\Psi)$$

We will assume the existence of such fixed points for every modal formula with one free variable. This is of course not an obvious thing to assume, but if we interpret $\Box$ as provability in Peano Arithmetic, then the existence of modal fixed points is in fact true.

Modal Rules of Inference

In addition to the existence of modal fixed points, we assume the following rules of inference for the provability operator $\Box$:

1. Necessitation: From $\vdash A$ conclude $\Box A$: Informally, this says that if A is a theorem, then it is provable.

2. Internal necessitation: $\vdash \Box A \rightarrow \Box\Box A$: If A is provable, then it is provable that it is provable.

3. Box distributivity: $\vdash \Box(A \rightarrow B) \rightarrow (\Box A \rightarrow \Box B)$: This rule allows you to do modus ponens inside the provability operator. If it is provable that A implies B, and A is provable, then B is provable.

Examples

An immediate corollary of Löb's theorem is that, if P is not provable in PA, then "if P is provable in PA, then P is true" is not provable in PA. Given we know PA is consistent (but PA does not know PA is consistent), here are some simple examples:

*   "If $1+1=3$ is provable in PA, then $1+1=3$" is not provable in PA, as $1+1=3$ is not provable in PA (as it is false).

*   "If $1+1=2$ is provable in PA, then $1+1=2$" is provable in PA, as is any statement of the form "If X, then $1+1=2$".

*   "If the strengthened finite Ramsey theorem is provable in PA, then the strengthened finite Ramsey theorem is true" is not provable in PA, as "The strengthened finite Ramsey theorem is true" is not provable in PA (despite being true).

In Doxastic logic, Löb's theorem shows that any system classified as a *reflexive* "type 4" reasoner must also be "*modest*": such a reasoner can never believe "my belief in P would imply that P is true", without first believing that P is true.

Converse: Löb's Theorem Implies the Existence of Modal Fixed Points

Not only does the existence of modal fixed points imply Löb's theorem, but the converse is valid, too. When Löb's theorem is given as an axiom (schema), the existence of a fixed point (up to provable equivalence) $p \leftrightarrow A(p)$ for any formula $A(p)$ *modalized in p* can be derived. Thus in normal modal logic, Löb's axiom is equivalent to the conjunction of the axiom schema 4, ($\Box A \to \Box\Box A$,) and the existence of modal fixed points.

Morley's Categoricity Theorem

A theory is called κ-categorical, or categorical in power κ, if it has one model up to isomorphism of cardinality κ. Morley's Categoricity Theorem states that if a theory of first order logic is categorical in some uncountable power κ, then it is categorical in every uncountable power. This striking result still provides the impetus and motivation for various areas of contemporary research within model theory.

The categoricity theorem is remarkable for a number of reasons. The LöwenheimSkolem theorem tells us that every theory in a countable language with an infinite model has a model of any infinite cardinality. It is counterintuitive that such a restrictive structural property as categoricity in an uncountable power holds as models get very large. Furthermore, many examples of theories categorical in every uncountable power were known before the categoricity theorem was discovered, but the proofs of these facts relied on properties of the theories themselves. For example, the categoricity in uncountable powers of the theory of algebraically closed fields of characteristic p s(zero or prime) depends on facts about transcendence degree. The categoricity theorem manages to prove that theories categorical in some uncountable power must be categorical in every uncountable power using model-theoretic methods alone.

Saharon Shelah soon became, and still is, the major driving force.

Let $\kappa, \lambda > \omega$.

If T is κ-categorical, then T is λ-categorical.

Lemma 1

If T is κ-categorical,

then T is totally transcendental and admits no Vaught pairs.

Lemma 2

If T is totally transcendental and admits no Vaught pairs,

then T is λ-categorical.

Lemma 1 – Total Transcendentality

T is totally transcendental if there is no tree of formulas with parameters that has inconsistent branching but consistent branches.

Fact: For any T there are arbitrarily big models $M \models T$ such that for every countable $B \subseteq M$, M realises only countably many types over B.(Tricky; involves Skolem hull of an indiscernible sequence.)

If T is κ-categorical, the κ-sized model must have this property. With Löwenheim-Skolem it follows that T is ω-stable:

Over any countable set, T has only countably many types.

Total transcendentality now follows easily.

Fact: Totally transcendental theories have prime models, and these are even very well behaved.

Lemma 1 – No Vaught Pairs

A Vaught pair is $M \prec N$ such that $M \neq N$ but for some formula $\varphi(x)$ (with parameters in M) we have $\varphi^M = \varphi^N$.

Vaught's Two-Cardinal Theorem: If there is a Vaught pair, then there is a Vaught pair $M \prec N$ such that M is countable and N is uncountable.

If T is totally transcendental, then we can even make N as big as we want. (A bit tricky; use prime models to build a chain of Vaught extensions.)

There is always a k-sized model N in which every formula with parameters has finitely many or k solutions. (Simple union of chains argument.)

It follows that k-categorical T cannot have a Vaught pair $M \prec N$ with $|M| = \omega, |N| = \kappa$, so it cannot have any Vaught pair.

Lemma 2 – Strongly Minimal Formula

A formula $\mu(x)$ is minimal in M if it has infinitely many solutions, but for every $\varphi(x)$ with parameters in M, either $\mu(x) \wedge \varphi(x)$ *or* $\mu(x) \wedge \neg\varphi(x)$ has only finitely many solutions. ($\rightarrow$ *generic type*).

If T is totally transcendental and $M \models T$, there is a minimal formula in M.

(The formula may have parameters in M, but this is only a superficial problem because we can take M to be a prime model of T. Let's assume it has no parameters.)

If T has no Vaught pairs, every minimal formula is strongly minimal. (Proof uses elimination of $\exists^\infty$.)

Lemma 2 – Matroid

A matroid (M, cl) (a.k.a. pregeometry) is a set M together with a finitary closure operator cl on M which satisfies the exchange law:

$$a \in cl(C \cup \{b\}) \setminus cl\, C \implies b \in cl(C \cup \{a\}).$$

Examples: Linear hull in a vector space, algebraic closure in a field, graphical matroid.

Generating set: cl $A = M$. Independent set: $a \notin cl(A \setminus \{a\})$. Basis = minimal generating set = maximal independent set.

All bases have the same cardinality: rank/dimension of the matroid. $\mu(x)$ is a strongly minimal formula, then (μ^M, acl) is a matroid. Moreover, any two independent sequences have the same type.

(The converse is also true.)

Lemma 2 – Proving λ-Categoricity

Let $N_1, N_2 \models T$ such that $|N_1| = |N_2| = \lambda$.

By Löwenheim-Skolem and non-existence of Vaught pairs,

$$|\mu^{N_i}| = |N_i| = \lambda. \text{ Let } B_i \subseteq \mu^{N_i} \text{ be a basis of } \mu^{N_i}. \text{ Then } |B_i| = |\mu^{N_i}| = \lambda.$$

So B_1 and B_2 have the same type, i.e. there is a partial isomorphism between them. It extends to a partial isomorphism between μ^{N_i} and μ^{N_2}.

By existence of prime models and non-existence of Vaught pairs, N_1 is prime over μ^{N_i}. The partial isomorphism therefore extends to an embedding of N_1 into N_2. By non-existence of Vaught pairs, this must be an isomorphism $N_1 \simeq N_2$.

Examples of theories categorical in an uncountable power are somewhat limited. However, some natural examples of such theories include:

- Algebraically closed fields of characteristic p (zero or prime).
- Pure identity theory.
- Torsion-free divisible Abelian groups.
- Infinite Abelian groups in which every element has order p (prime).
- Natural numbers with a successor function.
- Vector spaces over a countable field.

Löwenheim–Skolem Theorem

The Löwenheim-Skolem theorem is a basic result in the *model theory* of first-order logic and is part of a family of closely related theorems that concern the relation between structures or models of first-order theories of different cardinality.

One idea behind it is that finitary first-order logic by itself cannot pin down the cardinality of infinite structures or models: roughly speaking, Löwenheim-Skolem says that given a such structure MM, one can construct elementary substructures or elementary extensions NN of different cardinalities. This gives an elementary reason why one cannot expect absolute categoricity for first-order theories.

The Löwenheim-Skolem Theorem says that if M is an infinite model in some language L, then for every cardinal $\kappa \geq |L|$, there is a model N of cardinality κ, elementarily equivalent to M.

More precisely, one has two theorems:

Downward Löwenheim-Skolem Theorem: Let M be an infinite model in some language L. Then for any subset $S \subseteq M$, there exists an elementary substructure $N \prec M$ containing S, with $|N| = |S| + |L|$. In particular, taking S to be an arbitrary subset of size κ with $|L| \le \kappa \le |M|$, , we can find an elementary substructure of M of size κ.

Proof of Downward Löwenheim-Skolem Theorem

Let M be a structure. For each non-empty definable subset D of M, choose some element $e(D) \in D$, using the axiom of choice. If X is any subset of M, let

$$c(X) = X \cup \{e(D) : D \text{ definable over } X, D \ne \theta\}$$

Note that over a set of size λ, there are at most $\lambda + |L|$ definable sets. Consequently,

$$|c(X)| \le |X| + |L|$$

Now given $S \subseteq M$ as in the theorem, let

$$N = S \cup c(S) \cup c(c(S)) \cup \cdots$$

By basic cardinal arithmetic, $|N| = |S| + |L|$. Then $N \prec M$ by the Tarski-Vaught test. Indeed, if D is a subset of M definable over N, then D uses only finitely many parameters, and is therefore definable over $c^{(i)}(S) \subseteq N$ for some i. Then

$$e(D) \in c^{(i+1)}(S) \subset N,$$

so $e(D)$ is an element of $N \cap D$ s. Therefore, every non-empty N-definable set intersects N. Therefore the Tarski-Vaught criterion holds and N is an elementary substructure of M. It has the correct size.

Upward Löwenheim-Skolem Theorem: Let M be an infinite model in some language L. Then for every cardinal κ bigger than and $|L|$ |, there is an elementary extension of M of size κ.

Proof of Upward Löwenheim-Skolem Theorem

Given an infinite structure M and a cardinal κ at least as big as both $|M|$ and $|L|$, let T be the union of the elementary diagram of M and the collection of statements:

$$\{c\alpha \ne c\beta : \alpha < \beta < \kappa\}$$

where $\{c\alpha\}_{\alpha < \kappa}$ is a collection of κ new constant symbols. By compactness, T is consistent. Indeed, any finite subset of T only mentions finitely many of the $c\alpha$ and therefore has a model consisting of M with the finitely many $c\alpha$ interpreted as distinct elements of M. So by compactness we can find a model $N \vDash T$. Then N is a model of the elementary diagram of M, so N is an elementary extension of M. Also, the $c\alpha$ ensure that N contains at least κ distinct elements, i.e., $|N| \ge \kappa$. There is a possiblity that N is too big; to hit κ on the nose, we use Downward Löwenheim-Skolem to find an elementary substructure of N having size κ and containing M. On general grounds, the resulting structure is an elementary extension of M.

On the level of theories, the Löwenheim-Skolem Theorem implies that if T is a theory with an infinite model, then T has a model of cardinality κ for every infinite $\kappa \geq |T|$.

These statements become slightly simpler when working in a countable language. In this case, Upward Löwenheim-Skolem says that if M is an infinite structure, then M has elementary extensions of all cardinalities greater than $|M|$. Similarly, Downward Löwenheim-Skolem implies that if M is an infinite structure, then M has elementary substructures of all infinite sizes less than $|M|$.

Examples and Consequences

Let **N** denote the natural numbers and R the reals. It follows from the theorem that the theory of $(\mathbf{N}, +, \times, 0, 1)$ (the theory of true first-order arithmetic) has uncountable models, and that the theory of $(\mathbf{R}, +, \times, 0, 1)$ (the theory of real closed fields) has a countable model. There are, of course, axiomatizations characterizing $(\mathbf{N}, +, \times, 0, 1)$ and $(\mathbf{R}, +, \times, 0, 1)$ up to isomorphism. The Löwenheim–Skolem theorem shows that these axiomatizations cannot be first-order. For example, the completeness of a linear order, which is used to characterize the real numbers as a complete ordered field, is a non-first-order property.

A theory is called categorical if it has only one model, up to isomorphism. This term was introduced by Veblen (1904), and for some time thereafter mathematicians hoped they could put mathematics on a solid foundation by describing a categorical first-order theory of some version of set theory. The Löwenheim–Skolem theorem dealt a first blow to this hope, as it implies that a first-order theory which has an infinite model cannot be categorical. Later, in 1931, the hope was shattered completely by Gödel's incompleteness theorem.

Many consequences of the Löwenheim–Skolem theorem seemed counterintuitive to logicians in the early 20th century, as the distinction between first-order and non-first-order properties was not yet understood. One such consequence is the existence of uncountable models of true arithmetic, which satisfy every first-order induction axiom but have non-inductive subsets. Another consequence that was considered particularly troubling is the existence of a countable model of set theory, which nevertheless must satisfy the sentence saying the real numbers are uncountable. This counterintuitive situation came to be known as Skolem's paradox; it shows that the notion of countability is not absolute.

Lindström's Theorem

Lindström's Theorem is a model-theoretic characterization of first order logic. It says in effect that any formal language that goes beyond first order logic has to distinguish between some infinite cardinalities in the sense that some sentence has a model of some infinite cardinality but not of all infinite cardinalities. Loosely speaking Lindström's Theorem tells us that any proper extension of first order logic has to detect something non-trivial about the set-theoretic universe. The equivalent original formulation says that first order logic is a maximal logic which satisfies the Downward Löwenheim Skolem Property and the Countable Compactness Property.

As a first approximation, an abstract logic is a pair $L = (S, T)$, where S is a set and T is a relation between arbitrary structures and elements of the set S. Intuitively, S is the set of sentences of the abstract logic L and T is the truth predicate. If τ is a vocabulary, let $\text{Str}(\tau)$ denote the class of structures of vocabulary τ. Obviously we make some assumptions about S and T. They are listed below after we introduce some new concepts:

For $\varphi \in S$ we write $Mod_{L,\tau}(\varphi) = \{\mathfrak{M} \in Str(\tau) : T(\mathfrak{M}, \varphi)\}$. An abstract logic L is said to be closed under negation, if for all vocabularies τ and all $\varphi \in S$ there is $\neg\varphi \in S$ such that $Mod_{L,\tau}(\neg\varphi) = Str(\tau) \setminus Mod_{L,\tau}(\varphi)$. We say L is closed under conjunction if for all vocabularies τ and all $\varphi, \psi \in S$ there is $\varphi \wedge \in 2\,S$ such that $Mod_{L,\tau}(\varphi \wedge \psi) = Mod_{L,\tau}(\varphi) \cap Mod_{L,\tau}(\psi)$. We say L is closed under existential quantification, if for all vocabularies τ, for all constant symbols c in τ and for all $\varphi \in S$, there is $\varphi' \in S$ such that:

$$Mod_{L,\tau\setminus\{c\}}(\varphi') = \{\mathfrak{M} : (\mathfrak{M}, c^{\mathfrak{M}}) \in Mod_{L,\tau}(\varphi) \text{ for some } c^{\mathfrak{M}} \in M\}.$$

We say that L is closed under renaming if whenever $\pi : \tau \to \tau'$ is a permutation which respects arity, and we extend π in a canonical way to $\hat{\pi} : Str(\tau) \to Str(\tau')$, then for all and $\varphi \in S$, there is $\varphi' \in S$ such that $\{\hat{\pi}(\mathfrak{M}) : \mathfrak{M} \in Mod_{L,\tau}(\varphi)\} = Mod_{L,\tau'}(\varphi')$. We say that L is closed under free expansions if whenever $\tau \subseteq \tau'$ and $\varphi \in S$, there is $\varphi' \in$ such that $Mod_{L,\tau}(\varphi) = Mod_{L,\tau'}(\varphi')$. Finally, we say that L is closed under isomorphisms, if whenever $\varphi \in S$, $\mathfrak{M} \in Mod_{L,\tau}(\varphi)$ and $f : \mathfrak{M} \cong \mathfrak{N}$, then also $\mathfrak{N} \in Mod_{L,\tau}(\varphi)$.

We get an abstract logic $L_{\omega\omega} = (S_0, T_0)$ satisfying the above closure properties by letting S_0 be the set of all first order sentences and To the usual truth predicate of first order logic:

$$T_0(\mathfrak{M}, \varphi) \Leftrightarrow \mathfrak{M} \models \varphi$$

For example, the closure under free expansions can be satisfied simply by choosing $\varphi' = \varphi$. Other abstract logics arise from infinitary languages, generalized quantifiers, higher order logic and combinations of such.

An abstract logic $L = (S, T)$ is a sublogic of another abstract logic $L' = (S', T')$, in symbols $L \le L'$, if for all $\varphi \in S$ there is $\varphi' \in S'$ such that for all τ $Mod_{L,\tau}(\varphi) = Mod_{L,\tau'}(\varphi')$.. If $L \le L'$ and $L \le L'$, we say that L and L' are equivalent, $L \equiv L'$.

Now we are ready to give the real definition:

An abstract logic is a pair $L = (S, T)$, where S is a set and T is a relation between structures and elements of S, such that L is closed under isomorphisms, renaming, free expansions, negation, conjunction, and existential quantification.

An abstract logic $L = (S, T)$ satisfies the (Countable) Compactness Property if for any (countable) $\Sigma \subseteq S$, if $\cap\{Mod_{M,\tau}(\varphi) : \varphi \in \Sigma\} = \theta$, then $\cap\{Mod_{M,\tau}(\varphi) : \varphi \in \Sigma_0\} = \theta$, for some finite $\Sigma_0 \subseteq \Sigma$. An abstract logic $L = (S, T)$ satisfies the Downward Löwenheim-Skolem Property if for every countable τ every non-empty $Mod_{L,\tau}(\varphi)$, $\varphi \in S$, contains a countable model. Now we are ready to state Lindström's Theorem:

Lindström's Theorem: Suppose L is an abstract logic such that $FO \leq L$. Then the following conditions are equivalent:

(1) L has the Countable Compactness Property and the Downward Löwenheim-Skolem Property.

(2) $L \equiv L_{\omega\omega}$.

There are several other characterizations of first order logic, all due to Lindström. An abstract logic $L = (S, T)$ satisfies the Upward Löwenheim Skolem Property if every $Mod_{L,\tau}(\varphi), \varphi \in S$, which contains an infinite model, contains an uncountable model. We can replace condition (1) above by

(1)' L has the Upward and Downward Löwenheim-Skolem Properties.

In this form Lindström's characterization of first order logic among all abstract logics extends Mostowski's characterization of $L_{\omega\omega}$ among logics obtained by adding simple unary generalized quantifiers to $L_{\omega\omega}$.

For another characterization, we fix some notation. We use $Th_L(\mathfrak{M})$ to denote $\{\varphi \in S : T(\mathfrak{M}, \varphi)\}$. If $\mathfrak{M} \subseteq \mathfrak{N}$ and $Th_L(\mathfrak{M}(, a)_{a \in \mathfrak{M}}) = Th_L((\mathfrak{N}, a)a \in N)$,

we write $\mathfrak{M} <_L \mathfrak{N}$. An abstract logic $L = (S, T)$ satisfies the Tarski Union Property if $\mathfrak{M}_0 <_L \mathfrak{M}_1 <_L \dots$ implies $\mathfrak{M}_n <_L \bigcup_n \mathfrak{M}_n$. We can replace condition (1) above by

(1)" L has the Compactness Property and the Tarski Union Property.

We can also replace (1) by a condition derived from the Omitting-Types Theorem of first order logic. By assuming a little more effectiveness about the abstract logics, (1) can be replaced by a combination of the Downward Löwenheim-Skolem Property and either a property derived from the Beth Definability Theorem of first order logic, extending a result of Mostowski, or a property derived from the Completeness Theorem of first order logic. Lindström has himself written a very readable survey of his results in.

Barwise characterized $L_{\kappa\omega}$ for $\kappa = \beth_\kappa$. Lindström's Theorem was rediscovered later by H. Friedman.

Proof of Lindström's Theorem

Suppose there were an abstract logic $L = (S, T)$ that satisfies both the Downward Löwenheim Property and the Countable Compactness Property, but some $\varphi \in S$ is not first order definable, i.e. $Mod_{L,\tau}(\varphi)$ is not of the form $Mod_{L_{\omega\omega},\tau}(\psi)$ for any first order (ψ). We assume w.l.o.g. that is τ finite and relational.

For every n there are only finitely many (logically non-equivalent) first order sentences ψ_i^n, $i = 1, \dots, k_n$, of vocabulary τ and of quantifier rank at most n. Let us call two L-structures n-equivalent if they satisfy the same ψ_i^n. There are only $\leq 2^{k_n}$ different n-equivalence classes, and each class is first order definable. Since φ is not definable in first order logic, we can find for any n L-structures   and $\mathfrak{N}_n$ such that:

$$T(\mathfrak{M}_n, \varphi)$$

$$T\left(\mathfrak{N}_n, \neg\varphi\right)$$

$\mathfrak{M}_n$ and $\mathfrak{N}_n$ are n-equivalent.

Lindström uses then a characterization of n-equivalence in terms of backand-forth sequences. Ehrenfeucht and Fraisse showed that two models are n-equivalent if and only if there are relations $I_i, i < n$, such that

If $\left(a_1, ..., a_i\right)I_i\left(b_1, ..., b_i\right)$, then $a_1, ..., a_i \in M$ and $b_1, ..., b_i \in N$

- $() I_0 ()$

- If $\left(a_1, ..., a_i\right)I_i\left(b_1, ..., b_i\right)$ then for all $a_{i+1} \in M$ $\left(b_{i+1} \in N\right)$ there is $b_{i+1} \in N$ $\left(a_{i+1} \in M\right)$ such that $\left(a_1, ..., a_{i+1}\right)I_{i+1}\left(b_1, ..., b_{i+1}\right)$.

- If $\left(a_1, ..., a_{i-1}\right)I_i\left(b_1, ..., b_{i-1}\right)$, then for all atomic formulas $\varphi\left(v_1, ..., v_{i-1}\right)$ we have $\mathfrak{M} \models \varphi\left(a_1, ..., a_{i-1}\right)$ if and only if $\mathfrak{N} \models \varphi\left(b_1, ..., b_{i-1}\right)$.

If there are such relations I_i, $i < \omega$, then we say that that $\mathfrak{M}$ and $\mathfrak{N}$ are ω-equivalent.

If $\mathfrak{M}$ and $\mathfrak{N}$ are countable and ω-equivalent, then $\mathfrak{M} \cong \mathfrak{N}$, as we can go "back-and-forth" between the countable models generating infinite sequences $\left(a_1, ..., a_i, ...\right)$ and $\left(b_1, ..., b_i, ...\right)$ such that for all i we have $\left(a_1, ..., a_i\right)I_i\left(b_1, ..., b_i\right)$, and moreover, $M = \left\{a_i : i < w\right\}$ and $N = \left\{b_i : i < \omega\right\}$.

Lindström writes (1), supplemented with a little bit of arithmetic, into a sentence $\psi\left(n\right)$ in S, using the above back-and-forth characterization of n-equivalence. By the Countable Compactness Property there is a model of $\psi\left(n\right)$ in which n is non-standard. Due to the coding used, this model yields two other models $\mathfrak{M}$ and $\mathfrak{N}$ such that $T\left(\mathfrak{M}, \varphi\right)$, $T\left(\mathfrak{N}, \neg\varphi\right)$ and $\mathfrak{M}$ and $\mathfrak{N}$ are ω-equivalent. By the Downward Löwenheim-Skolem Property, we may assume $\mathfrak{M}$ and $\mathfrak{N}$ are countable. But then they are isomorphic by (2). Thus L cannot be closed under isomorphisms, contrary to assumption.

References

- Brady, Geraldine (2000), From Peirce to Skolem: A Neglected Chapter in the History of Logic, Elsevier, ISBN 978-0-444-50334-3

- Per Lindström (June 2006). "Note on Some Fixed Point Constructions in Provability Logic". Journal of Philosophical Logic. 35 (3): 225–230. doi:10.1007/s10992-005-9013-8

- Godel-completeness-theorem: terrytao.wordpress.com, Retrieved 31 March 2018

- Verbrugge, Rineke (L.C.) (1 January 2016). "Provability Logic". The Stanford Encyclopedia of Philosophy. Retrieved 6 April 2016

- Löwenheim, Leopold (1915), "Über Möglichkeiten im Relativkalkül" (PDF), Mathematische Annalen, 76 (4): 447–470, doi:10.1007/BF01458217, ISSN 0025-5831

- Goedel-incompleteness, entries: plato.stanford.edu, Retrieved 19 July 2018

- Poizat, Bruno (2000), A Course in Model Theory: An Introduction to Contemporary Mathematical Logic, Berlin, New York: Springer, ISBN 978-0-387-98655-5

- Löb, Martin (1955), "Solution of a Problem of Leon Henkin", Journal of Symbolic Logic, 20 (2): 115–118, JSTOR 2266895

Logical Systems

A logical system is generally characterized by the properties of consistency, validity, completeness and soundness. The aim of this chapter is to examine the different aspects of logical systems such as first-order, second-order and intuitionistic logic. It also elaborates axiomatic and Hilbert systems.

Logic is often studied by constructing what are commonly called logical systems. A logical system is essentially a way of mechanically listing all the logical truths of some part of logic by means of the application of recursive rules—i.e., rules that can be repeatedly applied to their own output. This is done by identifying by purely formal criteria certain axioms and certain purely formal rules of inference from which theorems can be derived from axioms together with earlier theorems. All of the axioms must be logical truths, and the rules of inference must preserve logical truth. If these requirements are satisfied, it follows that all the theorems in the system are logically true. If all the truths of the relevant part of logic can be captured in this way, the system is said to be "complete" in one sense of this ambiguous term.

The systematic study of formal derivations of logical truths from the axioms of a formal system is known as proof theory. It is one of the main areas of systematic logical theory.

Not all parts of logic are completely axiomatizable. Second-order logic, for example, is not axiomatizable on its most natural interpretation. Likewise, independence-friendly first-order logic is not completely axiomatizable. Hence the study of logic cannot be restricted to the axiomatization of different logical systems. One must also consider their semantics, or the relations between sentences in the logical system and the structures (usually referred to as "models") in which the sentences are true.

Logical systems that are incomplete in the sense of not being axiomatizable can nevertheless be formulated and studied in ways other than by mechanically listing all their logical truths. The notions of logical truth and validity can be defined model-theoretically (i.e., semantically) and studied systematically on the basis of such definitions without referring to any logical system or to any rules of inference. Such studies belong to model theory, which is another main branch of contemporary logic.

Model theory involves a notion of completeness and incompleteness that differs from axiomatizability. A system that is incomplete in the latter sense can nevertheless be complete in the sense that all the relevant logical truths are valid model-theoretical consequences of the system. This kind of completeness, known as descriptive completeness, is also sometimes (confusingly) called axiomatizability, despite the more common use of this term to refer to the mechanical generation of theorems from axioms and rules of inference.

First-order Logic

First-order logic is symbolized reasoning in which each sentence, or statement, is broken down into a subject and a predicate. The predicate modifies or defines the properties of the subject. In first-order logic, a predicate can only refer to a single subject. First-order logic is also known as first-order predicate calculus or first-order functional calculus.

A sentence in first-order logic is written in the form Px or P(x), where P is the predicate and x is the subject, represented as a variable. Complete sentences are logically combined and manipulated according to the same rules as those used in Boolean algebra.

In first-order logic, a sentence can be structured using the universal quantifier (symbolized $\forall$) or the existential quantifier ($\exists$). Consider a subject that is a variable represented by x. Let A be a predicate "is an apple," F be a predicate "is a fruit," S be a predicate "is sour'", and M be a predicate "is mushy." Then we can say

$$\forall x : Ax \Rightarrow Fx$$

which translates to "For all x, if x is an apple, then x is a fruit." We can also say such things as

$$\exists x : Fx \Rightarrow Ax$$

$$\exists x : Ax \Rightarrow Sx$$

$$\exists x : Ax \Rightarrow Mx$$

where the existential quantifier translates as "For some."

First-order logic can be useful in the creation of computer programs. It is also of interest to researchers in artificial intelligence (AI). There are more powerful forms of logic, but first-order logic is adequate for most everyday reasoning. The Incompleteness Theorem , proven in 1930, demonstrates that first-order logic is in general undecidable. That means there exist statements in this logic form that, under certain conditions, cannot be proven either true or false.

Syntax

There are two key parts of first-order logic. The syntax determines which collections of symbols are legal expressions in first-order logic, while the semantics determine the meanings behind these expressions.

Alphabet

Unlike natural languages, such as English, the language of first-order logic is completely formal, so that it can be mechanically determined whether a given expression is legal. There are two key types of legal expressions: terms, which intuitively represent objects, and formulas, which intuitively express predicates that can be true or false. The terms and formulas of first-order logic are strings of symbols, where all the symbols together form the alphabet of the language. As with all formal languages, the nature of the symbols themselves is outside the scope of formal logic; they are often regarded simply as letters and punctuation symbols.

It is common to divide the symbols of the alphabet into logical symbols, which always have the same meaning, and non-logical symbols, whose meaning varies by interpretation. For example, the logical symbol always represents "and"; it is never interpreted as "or". On the other hand, a non-logical predicate symbol such as Phil(x) could be interpreted to mean "x is a philosopher", "x is a man named Philip", or any other unary predicate, depending on the interpretation at hand.

Logical Symbols

There are several logical symbols in the alphabet, which vary by author but usually include:

- The quantifier symbols ∀ and ∃

- The logical connectives: ∧ for conjunction, ∨ for disjunction, → for implication, ↔ for biconditional, ¬ for negation. Occasionally other logical connective symbols are included. Some authors use Cpq, instead of →, and Epq, instead of ↔, especially in contexts where → is used for other purposes. Moreover, the horseshoe ⊃ may replace →; the triple-bar ≡ may replace ↔; a tilde (~), Np, or Fpq, may replace ¬; ||, or Apq may replace ∨; and &, Kpq, or the middle dot, ·, may replace ∧, especially if these symbols are not available for technical reasons. (*Note*: the aforementioned symbols Cpq, Epq, Np, Apq, and Kpq are used in Polish notation.)

- Parentheses, brackets, and other punctuation symbols. The choice of such symbols varies depending on context.

- An infinite set of variables, often denoted by lowercase letters at the end of the alphabet x, y, z, ... Subscripts are often used to distinguish variables: x_0, x_1, x_2, ...

- An equality symbol (sometimes, identity symbol) =.

It should be noted that not all of these symbols are required – only one of the quantifiers, negation and conjunction, variables, brackets and equality suffice. There are numerous minor variations that may define additional logical symbols:

- Sometimes the truth constants T, Vpq, or ⊤, for "true" and F, Opq, or ⊥, for "false" are included. Without any such logical operators of valence 0, these two constants can only be expressed using quantifiers.

- Sometimes additional logical connectives are included, such as the Sheffer stroke, Dpq (NAND), and exclusive or, Jpq.

Non-logical Symbols

The non-logical symbols represent predicates (relations), functions and constants on the domain of discourse. It used to be standard practice to use a fixed, infinite set of non-logical symbols for all purposes. A more recent practice is to use different non-logical symbols according to the application one has in mind. Therefore, it has become necessary to name the set of all non-logical symbols used in a particular application. This choice is made via a signature.

The traditional approach is to have only one, infinite, set of non-logical symbols (one signature) for all applications. Consequently, under the traditional approach there is only one language of first-order logic. This approach is still common, especially in philosophically oriented books.

1. For every integer $n \geq 0$ there is a collection of n-ary, or n-place, predicate symbols. Because they represent relations between n elements, they are also called relation symbols. For each arity n we have an infinite supply of them:

    $P^n_0, P^n_1, P^n_2, P^n_3, \dots$

2. For every integer $n \geq 0$ there are infinitely many n-ary function symbols:

    $f^n_0, f^n_1, f^n_2, f^n_3, \dots$

In contemporary mathematical logic, the signature varies by application. Typical signatures in mathematics are $\{1, \times\}$ or just $\{\times\}$ for groups, or $\{0, 1, +, \times, <\}$ for ordered fields. There are no restrictions on the number of non-logical symbols. The signature can be empty, finite, or infinite, even uncountable. Uncountable signatures occur for example in modern proofs of the Löwenheim–Skolem theorem.

In this approach, every non-logical symbol is of one of the following types.

1. A predicate symbol (or relation symbol) with some valence (or arity, number of arguments) greater than or equal to 0. These are often denoted by uppercase letters P, Q, R,… .

    *   Relations of valence 0 can be identified with propositional variables. For example, P, which can stand for any statement.

    *   For example, $P(x)$ is a predicate variable of valence 1. One possible interpretation is "x is a man".

    * $Q(x,y)$ is a predicate variable of valence 2. Possible interpretations include "x is greater than y" and "x is the father of y".

2. A function symbol, with some valence greater than or equal to 0. These are often denoted by lowercase letters f, g, h,… .

    *   Examples: $f(x)$ may be interpreted as for "the father of x". In arithmetic, it may stand for "$-x$". In set theory, it may stand for "the power set of x". In arithmetic, $g(x,y)$ may stand for "$x+y$". In set theory, it may stand for "the union of x and y".

    *   Function symbols of valence 0 are called constant symbols, and are often denoted by lowercase letters at the beginning of the alphabet a, b, c,… . The symbol a may stand for Socrates. In arithmetic, it may stand for 0. In set theory, such a constant may stand for the empty set.

The traditional approach can be recovered in the modern approach by simply specifying the "custom" signature to consist of the traditional sequences of non-logical symbols.

Formation Rules

The formation rules define the terms and formulas of first order logic. When terms and formulas are represented as strings of symbols, these rules can be used to write a formal grammar for terms and formulas. These rules are generally context-free (each production has a single symbol on the left side), except that the set of symbols may be allowed to be infinite and there may be many start symbols, for example the variables in the case of terms.

Terms

The set of terms is inductively defined by the following rules:

1. Variables: Any variable is a term.

2. Functions: Any expression $f(t_1,...,t_n)$ of n arguments (where each argument t_i is a term and f is a function symbol of valence n) is a term. In particular, symbols denoting individual constants are nullary function symbols, and are thus terms.

Only expressions which can be obtained by finitely many applications of rules 1 and 2 are terms. For example, no expression involving a predicate symbol is a term.

Formulas

The set of formulas (also called well-formed formulas or WFFs) is inductively defined by the following rules:

1. Predicate symbols: If P is an n-ary predicate symbol and t_1, ..., t_n are terms then $P(t_1,...,t_n)$ is a formula.

2. Equality: If the equality symbol is considered part of logic, and t_1 and t_2 are terms, then $t_1 = t_2$ is a formula.

3. Negation: If φ is a formula, then $\neg\varphi$ is a formula.

4. Binary connectives: If φ and ψ are formulas, then $(\varphi \rightarrow \psi)$ is a formula. Similar rules apply to other binary logical connectives.

5. Quantifiers: If φ is a formula and x is a variable, then $\forall x\varphi$ (for all x, φ holds) and $\exists x\varphi$ (there exists x such that φ) are formulas.

Only expressions which can be obtained by finitely many applications of rules 1–5 are formulas. The formulas obtained from the first two rules are said to be atomic formulas.

For example,

$$\forall x\forall y(P(f(x)) \rightarrow \neg(P(x) \rightarrow Q(f(y),x,z)))$$

is a formula, if f is a unary function symbol, P a unary predicate symbol, and Q a ternary predicate symbol. On the other hand, $\forall xx \rightarrow$ is not a formula, although it is a string of symbols from the alphabet.

The role of the parentheses in the definition is to ensure that any formula can only be obtained in one way by following the inductive definition (in other words, there is a unique parse tree for each formula). This property is known as unique readability of formulas. There are many conventions for where parentheses are used in formulas. For example, some authors use colons or full stops instead of parentheses, or change the places in which parentheses are inserted. Each author's particular definition must be accompanied by a proof of unique readability.

This definition of a formula does not support defining an if-then-else function ite (c, a, b), where "c" is a condition expressed as a formula, that would return "a" if c is true, and "b" if it is false. This is because both predicates and functions can only accept terms as parameters, but the first parameter is a formula. Some languages built on first-order logic, such as SMT-LIB 2.0, add this.

Notational Conventions

For convenience, conventions have been developed about the precedence of the logical operators, to avoid the need to write parentheses in some cases. These rules are similar to the order of operations in arithmetic. A common convention is:

- $\neg$ is evaluated first
- $\wedge$ and $\vee$ are evaluated next
- Quantifiers are evaluated next
- $\rightarrow$ is evaluated last.

Moreover, extra punctuation not required by the definition may be inserted to make formulas easier to read. Thus the formula,

$$(\neg \forall x P(x) \rightarrow \exists x \neg P(x))$$

might be written as

$$(\neg [\forall x P(x)]) \rightarrow \exists x [\neg P(x)]$$

In some fields, it is common to use infix notation for binary relations and functions, instead of the prefix notation defined above. For example, in arithmetic, one typically writes "2 + 2 = 4" instead of "=(+(2,2),4)". It is common to regard formulas in infix notation as abbreviations for the corresponding formulas in prefix notation, also term structure vs. representation.

The definitions above use infix notation for binary connectives such as $\rightarrow$. A less common convention is Polish notation, in which one writes $\rightarrow$, $\wedge$, and so on in front of their arguments rather than between them. This convention allows all punctuation symbols to be discarded. Polish notation is compact and elegant, but rarely used in practice because it is hard for humans to read it. In Polish notation, the formula

$$\forall x \forall y (P(f(x)) \rightarrow \neg (P(x) \rightarrow Q(f(y), x, z)))$$

becomes "$\forall x \forall y \rightarrow Pfx \neg \rightarrow PxQfyxz$".

Free and Bound Variables

In a formula, a variable may occur free or bound. Intuitively, a variable is free in a formula if it is not quantified: in $\forall y\, P(x,y)$, variable x is free while y is bound. The free and bound variables of a formula are defined inductively as follows:

1. Atomic formulas: If φ is an atomic formula then x is free in φ if and only if x occurs in φ. Moreover, there are no bound variables in any atomic formula.

2. Negation: x is free in $\neg\varphi$ if and only if x is free in φ. x is bound in $\neg\varphi$ if and only if x is bound in φ.

3. Binary connectives: x is free in $(\varphi \rightarrow \psi)$ if and only if x is free in either φ or ψ. x is bound in $(\varphi \rightarrow \psi)$ if and only if x is bound in either φ or ψ. The same rule applies to any other binary connective in place of $\rightarrow$.

4. Quantifiers: x is free in $\forall y\, \varphi$ if and only if x is free in φ and x is a different symbol from y. Also, x is bound in $\forall y\, \varphi$ if and only if x is y or x is bound in φ. The same rule holds with $\exists$ in place of $\forall$.

For example, in $\forall x\, \forall y\, (P(x) \rightarrow Q(x, f(x), z))$, x and y are bound variables, z is a free variable, and w is neither because it does not occur in the formula.

Free and bound variables of a formula need not be disjoint sets: x is both free and bound in $P(x) \rightarrow \forall x Q(x)$.

Freeness and boundness can be also specialized to specific occurrences of variables in a formula. For example, in $P(x) \rightarrow \forall x Q(x)$, the first occurrence of x is free while the second is bound. In other words, the x in $P(x)$ is free while the x in $\forall x Q(x)$ is bound.

A formula in first-order logic with no free variables is called a first-order sentence. These are the formulas that will have well-defined truth values under an interpretation. For example, whether a formula such as $\text{Phil}(x)$ is true must depend on what x represents. But the sentence $\exists x\, \text{Phil}(x)$ will be either true or false in a given interpretation.

Example: Ordered Abelian Groups

In mathematics the language of ordered abelian groups has one constant symbol 0, one unary function symbol $-$, one binary function symbol $+$, and one binary relation symbol $\leq$. Then:

- The expressions $+(x, y)$ and $+(x, +(y, -(z)))$ are terms. These are usually written as $x + y$ and $x + y - z$.

- The expressions $+(x, y) = 0$ and $\leq(+(x, +(y, -(z))), +(x, y))$ are atomic formulas. These are usually written as $x + y = 0$ and $x + y - z \leq x + y$.

- The expression $(\forall x \forall y [\leq (+(x,y), z) \rightarrow \forall x \forall y + (x,y) = 0)]$ is a formula, which is usually written as $\forall x \forall y (x + y \leq z) \rightarrow \forall x \forall y (x + y = 0)$ This formula has one free variable, z.

The axioms for ordered abelian groups can be expressed as a set of sentences in the language. For example, the axiom stating that the group is commutative is usually written $(\forall x)(\forall y)[x + y = y + x]$.

Semantics

An interpretation of a first-order language assigns a denotation to all non-logical constants in that language. It also determines a domain of discourse that specifies the range of the quantifiers. The result is that each term is assigned an object that it represents, and each sentence is assigned a truth value. In this way, an interpretation provides semantic meaning to the terms and formulas of the language. The study of the interpretations of formal languages is called formal semantics. What follows is a description of the standard or Tarskian semantics for first-order logic. (It is also possible to define game semantics for first-order logic, but aside from requiring the axiom of choice, game semantics agree with Tarskian semantics for first-order logic, so game semantics will not be elaborated herein.)

The domain of discourse D is a nonempty set of "objects" of some kind. Intuitively, a first-order formula is a statement about these objects; for example, $\exists x P(x)$ states the existence of an object x such that the predicate P is true where referred to it. The domain of discourse is the set of considered objects. For example, one can take D to be the set of integer numbers.

The interpretation of a function symbol is a function. For example, if the domain of discourse consists of integers, a function symbol f of arity 2 can be interpreted as the function that gives the sum of its arguments. In other words, the symbol f is associated with the function $I(f)$ which, in this interpretation, is addition.

The interpretation of a constant symbol is a function from the one-element set D^0 to D, which can be simply identified with an object in D. For example, an interpretation may assign the value $I(c) = 10$ to the constant symbol c.

The interpretation of an n-ary predicate symbol is a set of n-tuples of elements of the domain of discourse. This means that, given an interpretation, a predicate symbol, and n elements of the domain of discourse, one can tell whether the predicate is true of those elements according to the given interpretation. For example, an interpretation $I(P)$ of a binary predicate symbol P may be the set of pairs of integers such that the first one is less than the second. According to this interpretation, the predicate P would be true if its first argument is less than the second.

First-order Structures

The most common way of specifying an interpretation (especially in mathematics) is to specify a structure. The structure consists of a nonempty set D that forms the domain of discourse and an interpretation I of the non-logical terms of the signature. This interpretation is itself a function:

- Each function symbol f of arity n is assigned a function $I(f)$ from D^n to D. In particular, each constant symbol of the signature is assigned an individual in the domain of discourse.

- Each predicate symbol P of arity n is assigned a relation $I(P)$ over D^n or, equivalently, a function from D^n to {true, false}. Thus each predicate symbol is interpreted by a Boolean-valued function on D.

Evaluation of Truth Values

A formula evaluates to true or false given an interpretation, and a variable assignment μ that associates an element of the domain of discourse with each variable. The reason that a variable assignment is required is to give meanings to formulas with free variables, such as $y = x$. The truth value of this formula changes depending on whether x and y denote the same individual.

First, the variable assignment μ can be extended to all terms of the language, with the result that each term maps to a single element of the domain of discourse. The following rules are used to make this assignment:

1.   Variables: Each variable x evaluates to $\mu(x)$

2.   Functions: Given terms $t_1,...,t_n$ that have been evaluated to elements $d_1,...,d_n$ of the domain of discourse, and a n-ary function symbol f, the term $f(t_1,...,t_n)$ evaluates to $(I(f))(d_1,...,d_n)$.

Next, each formula is assigned a truth value. The inductive definition used to make this assignment is called the T-schema.

1.   Atomic formulas (1): A formula $P(t_1,...,t_n)$ is associated the value true or false depending on whether $\langle v_1,...,v_n\rangle \in I(P)$, where $v_1,...,v_n$ are the evaluation of the terms $t_1,...,t_n$ and $I(P)$ is the interpretation of P, which by assumption is a subset of D^n.

2.   Atomic formulas (2): A formula $t_1 = t_2$ is assigned true if t_1 and t_2 evaluate to the same object of the domain of discourse.

3.   Logical connectives: A formula in the form $\neg\phi$, $\phi \rightarrow \psi$, etc. is evaluated according to the truth table for the connective in question, as in propositional logic.

4.   Existential quantifiers: A formula $\exists x\phi(x)$ is true according to M and μ if there exists an evaluation μ' of the variables that only differs from μ regarding the evaluation of x and such that φ is true according to the interpretation M and the variable assignment μ'. This formal definition captures the idea that $\exists x\phi(x)$ is true if and only if there is a way to choose a value for x such that $\varphi(x)$ is satisfied.

5.   Universal quantifiers: A formula $\forall x\phi(x)$ is true according to M and μ if $\varphi(x)$ is true for every pair composed by the interpretation M and some variable assignment μ' that differs from μ only on the value of x. This captures the idea that $\forall x\phi(x)$ is true if every possible choice of a value for x causes $\varphi(x)$ to be true.

If a formula does not contain free variables, and so is a sentence, then the initial variable assignment does not affect its truth value. In other words, a sentence is true according to M and μ if and only if it is true according to M and every other variable assignment μ'.

There is a second common approach to defining truth values that does not rely on variable assignment functions. Instead, given an interpretation M, one first adds to the signature a collection of constant symbols, one for each element of the domain of discourse in M; say that for each d in the domain the constant symbol c_d is fixed. The interpretation is extended so that each new constant symbol is assigned to its corresponding element of the domain. One now defines truth for quantified formulas syntactically, as follows:

1.   Existential quantifiers (alternate): A formula $\exists x\phi(x)$ is true according to M if there is some d in the domain of discourse such that $\phi(c_d)$ holds. Here $\phi(c_d)$ is the result of substituting c_d for every free occurrence of x in φ.

2.   Universal quantifiers (alternate): A formula $\forall x\phi(x)$ is true according to M if, for every d in the domain of discourse, $\phi(c_d)$ is true according to M.

This alternate approach gives exactly the same truth values to all sentences as the approach via variable assignments.

Validity, Satisfiability and Logical Consequence

If a sentence φ evaluates to True under a given interpretation M, one says that M satisfies φ; this is denoted $M \vDash \varphi$. A sentence is satisfiable if there is some interpretation under which it is true.

Satisfiability of formulas with free variables is more complicated, because an interpretation on its own does not determine the truth value of such a formula. The most common convention is that a formula with free variables is said to be satisfied by an interpretation if the formula remains true regardless which individuals from the domain of discourse are assigned to its free variables. This has the same effect as saying that a formula is satisfied if and only if its universal closure is satisfied.

A formula is logically valid (or simply valid) if it is true in every interpretation. These formulas play a role similar to tautologies in propositional logic.

A formula φ is a logical consequence of a formula ψ if every interpretation that makes ψ true also makes φ true. In this case one says that φ is logically implied by ψ.

Algebraizations

An alternate approach to the semantics of first-order logic proceeds via abstract algebra. This approach generalizes the Lindenbaum–Tarski algebras of propositional logic. There are three ways of eliminating quantified variables from first-order logic that do not involve replacing quantifiers with other variable binding term operators:

- Cylindric algebra, by Alfred Tarski and his coworkers;

- Polyadic algebra, by Paul Halmos;

- Predicate functor logic, mainly due to Willard Quine.

These algebras are all lattices that properly extend the two-element Boolean algebra.

Tarski and Givant (1987) showed that the fragment of first-order logic that has no atomic sentence lying in the scope of more than three quantifiers has the same expressive power as relation algebra. This fragment is of great interest because it suffices for Peano arithmetic and most axiomatic set theory, including the canonical ZFC. They also prove that first-order logic with a primitive ordered pair is equivalent to a relation algebra with two ordered pair projection functions.

First-order Theories, Models and Elementary Classes

A first-order theory of a particular signature is a set of axioms, which are sentences consisting of symbols from that signature. The set of axioms is often finite or recursively enumerable, in which case the theory is called effective. Some authors require theories to also include all logical consequences of the axioms. The axioms are considered to hold within the theory and from them other sentences that hold within the theory can be derived.

A first-order structure that satisfies all sentences in a given theory is said to be a model of the theory. An elementary class is the set of all structures satisfying a particular theory. These classes are a main subject of study in model theory.

Many theories have an intended interpretation, a certain model that is kept in mind when studying the theory. For example, the intended interpretation of Peano arithmetic consists of the usual natural numbers with their usual operations. However, the Löwenheim–Skolem theorem shows that most first-order theories will also have other, nonstandard models.

A theory is consistent if it is not possible to prove a contradiction from the axioms of the theory. A theory is complete if, for every formula in its signature, either that formula or its negation is a logical consequence of the axioms of the theory. Gödel's incompleteness theorem shows that effective first-order theories that include a sufficient portion of the theory of the natural numbers can never be both consistent and complete.

Empty Domains

The definition above requires that the domain of discourse of any interpretation must be a non-empty set. There are settings, such as inclusive logic, where empty domains are permitted. Moreover, if a class of algebraic structures includes an empty structure (for example, there is an empty poset), that class can only be an elementary class in first-order logic if empty domains are permitted or the empty structure is removed from the class.

There are several difficulties with empty domains, however:

- Many common rules of inference are only valid when the domain of discourse is required to be nonempty. One example is the rule stating that $\phi \lor \exists x \psi$ implies $\exists x (\phi \lor \psi)$ when x is not a free variable in ϕ. This rule, which is used to put formulas into prenex normal form, is sound in nonempty domains, but unsound if the empty domain is permitted.

- The definition of truth in an interpretation that uses a variable assignment function cannot work with empty domains, because there are no variable assignment functions whose range is empty. (Similarly, one cannot assign interpretations to constant symbols.) This truth definition requires that one must select a variable assignment function (μ above) before truth values for even atomic formulas can be defined. Then the truth value of a sentence is defined to be its truth value under any variable assignment, and it is proved that this truth value does not depend on which assignment is chosen. This technique does not work if there are no assignment functions at all; it must be changed to accommodate empty domains.

Thus, when the empty domain is permitted, it must often be treated as a special case. Most authors, however, simply exclude the empty domain by definition.

Deductive Systems

A deductive system is used to demonstrate, on a purely syntactic basis, that one formula is a logical consequence of another formula. There are many such systems for first-order logic, including Hilbert-style deductive systems, natural deduction, the sequent calculus, the tableaux method, and resolution. These share the common property that a deduction is a finite syntactic object; the format of this object, and the way it is constructed, vary widely. These finite deductions themselves are often called derivations in proof theory. They are also often called proofs, but are completely formalized unlike natural-language mathematical proofs.

A deductive system is sound if any formula that can be derived in the system is logically valid. Conversely, a deductive system is complete if every logically valid formula is derivable. All of the systems discussed in this topic are both sound and complete. They also share the property that it is possible to effectively verify that a purportedly valid deduction is actually a deduction; such deduction systems are called effective.

A key property of deductive systems is that they are purely syntactic, so that derivations can be verified without considering any interpretation. Thus a sound argument is correct in every possible interpretation of the language, regardless whether that interpretation is about mathematics, economics, or some other area.

In general, logical consequence in first-order logic is only semidecidable: if a sentence A logically implies a sentence B then this can be discovered (for example, by searching for a proof until one is found, using some effective, sound, complete proof system). However, if A does not logically imply B, this does not mean that A logically implies the negation of B. There is no effective procedure that, given formulas A and B, always correctly decides whether A logically implies B.

Rules of Inference

A rule of inference states that, given a particular formula (or set of formulas) with a certain property as a hypothesis, another specific formula (or set of formulas) can be derived as a conclusion. The rule is sound (or truth-preserving) if it preserves validity in the sense that whenever any interpretation satisfies the hypothesis, that interpretation also satisfies the conclusion.

For example, one common rule of inference is the rule of substitution. If t is a term and φ is a formula possibly containing the variable x, then $\varphi[t/x]$ is the result of replacing all free instances of x by t in φ. The substitution rule states that for any φ and any term t, one can conclude $\varphi[t/x]$ from φ provided that no free variable of t becomes bound during the substitution process. (If some free variable of t becomes bound, then to substitute t for x it is first necessary to change the bound variables of φ to differ from the free variables of t.)

To see why the restriction on bound variables is necessary, consider the logically valid formula φ given by $\exists x(x=y)$, in the signature of $(0,1,+,\times,=)$ of arithmetic. If t is the term "x + 1", the formula $\varphi[t/y]$ is $\exists x(x=x+1)$, which will be false in many interpretations. The problem is that the free variable x of t became bound during the substitution. The intended replacement can be obtained by renaming the bound variable x of φ to something else, say z, so that the formula after substitution is $\exists z(z=x+1)$, which is again logically valid.

The substitution rule demonstrates several common aspects of rules of inference. It is entirely syntactical; one can tell whether it was correctly applied without appeal to any interpretation. It has (syntactically defined) limitations on when it can be applied, which must be respected to preserve the correctness of derivations. Moreover, as is often the case, these limitations are necessary because of interactions between free and bound variables that occur during syntactic manipulations of the formulas involved in the inference rule.

Hilbert-style Systems and Natural Deduction

A deduction in a Hilbert-style deductive system is a list of formulas, each of which is a logical axiom, a hypothesis that has been assumed for the derivation at hand, or follows from previous formulas via a rule of inference. The logical axioms consist of several axiom schemas of logically valid formulas; these encompass a significant amount of propositional logic. The rules of inference enable the manipulation of quantifiers. Typical Hilbert-style systems have a small number of rules of inference, along with several infinite schemas of logical axioms. It is common to have only modus ponens and universal generalization as rules of inference.

Natural deduction systems resemble Hilbert-style systems in that a deduction is a finite list of formulas. However, natural deduction systems have no logical axioms; they compensate by adding additional rules of inference that can be used to manipulate the logical connectives in formulas in the proof.

Sequent Calculus

The sequent calculus was developed to study the properties of natural deduction systems. Instead of working with one formula at a time, it uses sequents, which are expressions of the form

$$A_1, \dots, A_n \vdash B_1, \dots, B_k$$

where $A_1, \dots, A_n, B_1, \dots, B_k$ are formulas and the turnstile symbol $\vdash$ is used as punctuation to separate the two halves. Intuitively, a sequent expresses the idea that $(A_1 \wedge \dots \wedge A_n)$ implies $(B_1 \vee \dots \vee B_k)$.

Tableaux Method

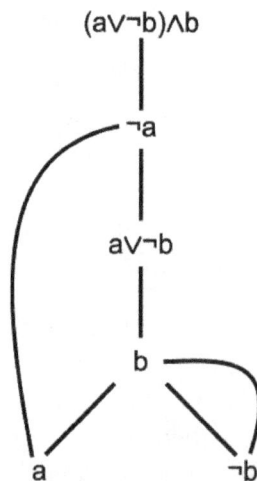

A tableaux proof for the propositional formula $((a \vee \neg b) \wedge b) \to a$.

Unlike the methods just described, the derivations in the tableaux method are not lists of formulas. Instead, a derivation is a tree of formulas. To show that a formula A is provable, the tableaux method attempts to demonstrate that the negation of A is unsatisfiable. The tree of the derivation has $\neg A$ at its root; the tree branches in a way that reflects the structure of the formula. For example, to show that $C \vee D$ is unsatisfiable requires showing that C and D are each unsatisfiable; this corresponds to a branching point in the tree with parent $C \vee D$ and children C and D.

Resolution

The resolution rule is a single rule of inference that, together with unification, is sound and complete for first-order logic. As with the tableaux method, a formula is proved by showing that the negation of the formula is unsatisfiable. Resolution is commonly used in automated theorem proving.

The resolution method works only with formulas that are disjunctions of atomic formulas; arbitrary formulas must first be converted to this form through Skolemization. The resolution rule states that from the hypotheses $A_1 \vee \cdots \vee A_k \vee C$ and $B_1 \vee \cdots \vee B_l \vee \neg C$, the conclusion $A_1 \vee \cdots \vee A_k \vee B_1 \vee \cdots \vee B_l$ can be obtained.

Provable Identities

Many identities can be proved, which establish equivalences between particular formulas. These identities allow for rearranging formulas by moving quantifiers across other connectives, and are useful for putting formulas in prenex normal form. Some provable identities include:

$$\neg \forall x P(x) \Leftrightarrow \exists x \neg P(x)$$

$$\neg \exists x P(x) \Leftrightarrow \forall x \neg P(x)$$

$$\forall x \forall y P(x,y) \Leftrightarrow \forall y \forall x P(x,y)$$

$$\exists x \exists y P(x,y) \Leftrightarrow \exists y \exists x P(x,y)$$

$$\forall x P(x) \wedge \forall x Q(x) \Leftrightarrow \forall x (P(x) \wedge Q(x))$$

$$\exists x P(x) \vee \exists x Q(x) \Leftrightarrow \exists x (P(x) \vee Q(x))$$

$$P \wedge \exists x Q(x) \Leftrightarrow \exists x (P \wedge Q(x)) \text{ (where } x \text{ must not occur free in } P\text{)}$$

$$P \vee \forall x Q(x) \Leftrightarrow \forall x (P \vee Q(x)) \text{ (where } x \text{ must not occur free in } P\text{)}$$

Equality and its Axioms

There are several different conventions for using equality (or identity) in first-order logic. The most common convention, known as first-order logic with equality, includes the equality symbol as a primitive logical symbol which is always interpreted as the real equality relation between members of the domain of discourse, such that the "two" given members are the same member. This approach also adds certain axioms about equality to the deductive system employed. These equality axioms are:

1. Reflexivity: For each variable x, $x = x$.

2. Substitution for functions: For all variables x and y, and any function symbol f,

    $$x = y \rightarrow f(...,x,...) = f(...,y,...).$$

3. Substitution for formulas: For any variables x and y and any formula $\varphi(x)$, if φ' is obtained by replacing any number of free occurrences of x in φ with y, such that these remain free occurrences of y, then

    $$x = y \rightarrow (\varphi \rightarrow \varphi').$$

These are axiom schemas, each of which specifies an infinite set of axioms. The third schema is known as Leibniz's law, "the principle of substitutivity", "the indiscernibility of identicals", or "the replacement property". The second schema, involving the function symbol f, is (equivalent to) a special case of the third schema, using the formula

$$x = y \rightarrow (f(...,x,...) = z \rightarrow f(...,y,...) = z).$$

Many other properties of equality are consequences of the axioms above, for example:

1. Symmetry: If $x = y$ then $y = x$.

2. Transitivity: If $x = y$ and $y = z$ then $x = z$.

First-order Logic without Equality

An alternate approach considers the equality relation to be a non-logical symbol. This convention is known as first-order logic without equality. If an equality relation is included in the signature, the axioms of equality must now be added to the theories under consideration, if desired, instead of being considered rules of logic. The main difference between this method and first-order logic with equality is that an interpretation may now interpret two distinct individuals as "equal" (although, by Leibniz's law, these will satisfy exactly the same formulas under any interpretation). That is, the equality relation may now be interpreted by an arbitrary equivalence relation on the domain of discourse that is congruent with respect to the functions and relations of the interpretation.

When this second convention is followed, the term normal model is used to refer to an interpretation where no distinct individuals a and b satisfy $a = b$. In first-order logic with equality, only normal models are considered, and so there is no term for a model other than a normal model. When first-order logic without equality is studied, it is necessary to amend the statements of results such as the Löwenheim–Skolem theorem so that only normal models are considered.

First-order logic without equality is often employed in the context of second-order arithmetic and other higher-order theories of arithmetic, where the equality relation between sets of natural numbers is usually omitted.

Defining Equality within a Theory

If a theory has a binary formula $A(x,y)$ which satisfies reflexivity and Leibniz's law, the theory is said to have equality, or to be a theory with equality. The theory may not have all instances of the above schemas as axioms, but rather as derivable theorems. For example, in theories with no function symbols and a finite number of relations, it is possible to define equality in terms of the relations, by defining the two terms s and t to be equal if any relation is unchanged by changing s to t in any argument.

Some theories allow other *ad hoc* definitions of equality:

- In the theory of partial orders with one relation symbol $\leq$, one could define $s = t$ to be an abbreviation for $s \leq t \wedge t \leq s$.

- In set theory with one relation $\in$, one may define $s = t$ to be an abbreviation for $\forall x\,(s \in x \leftrightarrow t \in x) \wedge \forall x\,(x \in s \leftrightarrow x \in t)$. This definition of equality then automatically satisfies the axioms for equality. In this case, one should replace the usual axiom of extensionality, which can be stated as $\forall x \forall y [\forall z (z \in x \Leftrightarrow z \in y) \Rightarrow x = y]$, with an alternative formulation $\forall x \forall y [\forall z (z \in x \Leftrightarrow z \in y) \Rightarrow \forall z (x \in z \Leftrightarrow y \in z)]$, which says that if sets x and y have the same elements, then they also belong to the same sets.

Metalogical Properties

One motivation for the use of first-order logic, rather than higher-order logic, is that first-order logic has many metalogical properties that stronger logics do not have. These results concern general properties of first-order logic itself, rather than properties of individual theories. They provide fundamental tools for the construction of models of first-order theories.

Completeness and Undecidability

Gödel's completeness theorem, proved by Kurt Gödel in 1929, establishes that there are sound, complete, effective deductive systems for first-order logic, and thus the first-order logical consequence relation is captured by finite provability. Naively, the statement that a formula φ logically implies a formula ψ depends on every model of φ; these models will in general be of arbitrarily large cardinality, and so logical consequence cannot be effectively verified by checking every model. However, it is possible to enumerate all finite derivations and search for a derivation of ψ from φ. If ψ is logically implied by φ, such a derivation will eventually be found. Thus first-order logical consequence is semidecidable: it is possible to make an effective enumeration of all pairs of sentences (φ, ψ) such that ψ is a logical consequence of φ.

Unlike propositional logic, first-order logic is undecidable (although semidecidable), provided that the language has at least one predicate of arity at least 2 (other than equality). This means that there is no decision procedure that determines whether arbitrary formulas are logically valid. This result was established independently by Alonzo Church and Alan Turing in 1936 and 1937, respectively, giving a negative answer to the Entscheidungsproblem posed by David Hilbert in 1928. Their proofs demonstrate a connection between the unsolvability of the decision problem for first-order logic and the unsolvability of the halting problem.

There are systems weaker than full first-order logic for which the logical consequence relation is decidable. These include propositional logic and monadic predicate logic, which is first-order logic restricted to unary predicate symbols and no function symbols. Other logics with no function symbols which are decidable are the guarded fragment of first-order logic, as well as two-variable logic. The Bernays–Schönfinkel class of first-order formulas is also decidable. Decidable subsets of first-order logic are also studied in the framework of description logics.

The Löwenheim–Skolem Theorem

The Löwenheim–Skolem theorem shows that if a first-order theory of cardinality λ has an infinite model, then it has models of every infinite cardinality greater than or equal to λ. One of the earliest results in model theory, it implies that it is not possible to characterize countability or uncountability in a first-order language. That is, there is no first-order formula $\varphi(x)$ such that an arbitrary structure M satisfies φ if and only if the domain of discourse of M is countable (or, in the second case, uncountable).

The Löwenheim–Skolem theorem implies that infinite structures cannot be categorically axiomatized in first-order logic. For example, there is no first-order theory whose only model is the real line: any first-order theory with an infinite model also has a model of cardinality larger than the continuum. Since the real line is infinite, any theory satisfied by the real line is also satisfied by some nonstandard models. When the Löwenheim–Skolem theorem is applied to first-order set theories, the nonintuitive consequences are known as Skolem's paradox.

The Compactness Theorem

The compactness theorem states that a set of first-order sentences has a model if and only if every finite subset of it has a model. This implies that if a formula is a logical consequence of an infinite set of first-order axioms, then it is a logical consequence of some finite number of those axioms. This theorem was proved first by Kurt Gödel as a consequence of the completeness theorem, but many additional proofs have been obtained over time. It is a central tool in model theory, providing a fundamental method for constructing models.

The compactness theorem has a limiting effect on which collections of first-order structures are elementary classes. For example, the compactness theorem implies that any theory that has arbitrarily large finite models has an infinite model. Thus the class of all finite graphs is not an elementary class (the same holds for many other algebraic structures).

There are also more subtle limitations of first-order logic that are implied by the compactness theorem. For example, in computer science, many situations can be modeled as a directed graph of states (nodes) and connections (directed edges). Validating such a system may require showing that no "bad" state can be reached from any "good" state. Thus one seeks to determine if the good and bad states are in different connected components of the graph. However, the compactness theorem can be used to show that connected graphs are not an elementary class in first-order logic, and there is no formula $\varphi(x,y)$ of first-order logic, in the logic of graphs, that expresses the idea that there is a path from x to y. Connectedness can be expressed in second-order logic, however, but not with only existential set quantifiers, as Σ^1_1 also enjoys compactness.

Lindström's Theorem

Per Lindström showed that the metalogical properties just discussed actually characterize first-order logic in the sense that no stronger logic can also have those properties (Ebbinghaus and Flum 1994, Chapter XIII). Lindström defined a class of abstract logical systems, and a rigorous definition of the relative strength of a member of this class. He established two theorems for systems of this type:

- A logical system satisfying Lindström's definition that contains first-order logic and satisfies both the Löwenheim–Skolem theorem and the compactness theorem must be equivalent to first-order logic.

- A logical system satisfying Lindström's definition that has a semidecidable logical consequence relation and satisfies the Löwenheim–Skolem theorem must be equivalent to first-order logic.

Limitations

Although first-order logic is sufficient for formalizing much of mathematics, and is commonly used in computer science and other fields, it has certain limitations. These include limitations on its expressiveness and limitations of the fragments of natural languages that it can describe.

For instance, first-order logic is undecidable, meaning a sound, complete and terminating decision algorithm for provability is impossible. This has led to the study of interesting decidable fragments such as C_2, first-order logic with two variables and the counting quantifiers $\exists^{\geq n}$ and $\exists^{\leq n}$ (these quantifiers are, respectively, "there exists at least n" and "there exists at most n").

Expressiveness

The Löwenheim–Skolem theorem shows that if a first-order theory has any infinite model, then it has infinite models of every cardinality. In particular, no first-order theory with an infinite model can be categorical. Thus there is no first-order theory whose only model has the set of natural numbers as its domain, or whose only model has the set of real numbers as its domain. Many extensions of first-order logic, including infinitary logics and higher-order logics, are more expressive in the sense that they do permit categorical axiomatizations of the natural numbers or real numbers. This expressiveness comes at a metalogical cost, however: by Lindström's theorem, the compactness theorem and the downward Löwenheim–Skolem theorem cannot hold in any logic stronger than first-order.

Formalizing Natural Languages

First-order logic is able to formalize many simple quantifier constructions in natural language, such as "every person who lives in Perth lives in Australia". But there are many more complicated features of natural language that cannot be expressed in (single-sorted) first-order logic. "Any logical system which is appropriate as an instrument for the analysis of natural language needs a much richer structure than first-order predicate logic".

Type	Example	Comment
Quantification over properties	If John is self-satisfied, then there is at least one thing he has in common with Peter	Requires a quantifier over predicates, which cannot be implemented in single-sorted first-order logic: $Zj \rightarrow \exists X(Xj \wedge Xp)$
Quantification over properties	Santa Claus has all the attributes of a sadist	Requires quantifiers over predicates, which cannot be implemented in single-sorted first-order logic: $\forall X(\forall x(Sx \rightarrow Xx) \rightarrow Xs)$
Predicate adverbial	John is walking quickly	Cannot be analysed as $Wj \wedge Qj$; predicate adverbials are not the same kind of thing as second-order predicates such as colour
Relative adjective	Jumbo is a small elephant	Cannot be analysed as $Sj \wedge Ej$; predicate adjectives are not the same kind of thing as second-order predicates such as colour
Predicate adverbial modifier	John is walking very quickly	-
Relative adjective modifier	Jumbo is terribly small	An expression such as "terribly", when applied to a relative adjective such as "small", results in a new composite relative adjective "terribly small"
Prepositions	Mary is sitting next to John	The preposition "next to" when applied to "John" results in the predicate adverbial "next to John"

Restrictions, Extensions and Variations

There are many variations of first-order logic. Some of these are inessential in the sense that they merely change notation without affecting the semantics. Others change the expressive power more significantly, by extending the semantics through additional quantifiers or other new logical symbols. For example, infinitary logics permit formulas of infinite size, and modal logics add symbols for possibility and necessity.

Restricted Languages

First-order logic can be studied in languages with fewer logical symbols than were described above.

- Because $\exists x\phi(x)$ can be expressed as $\neg\forall x\neg\phi(x)$, and $\forall x\phi(x)$ can be expressed as $\neg\exists x\neg\phi(x)$, either of the two quantifiers $\exists$ and $\forall$ can be dropped.

- Since $\phi\vee\psi$ can be expressed as $\neg(\neg\phi\wedge\neg\psi)$ and $\phi\wedge\psi$ can be expressed as $\neg(\neg\phi\vee\neg\psi)$, either $\vee$ or $\wedge$ can be dropped. In other words, it is sufficient to have $\neg$ and $\vee$, or $\neg$ and $\wedge$, as the only logical connectives.

- Similarly, it is sufficient to have only $\neg$ and $\rightarrow$ as logical connectives, or to have only the Sheffer stroke (NAND) or the Peirce arrow (NOR) operator.

- It is possible to entirely avoid function symbols and constant symbols, rewriting them via predicate symbols in an appropriate way. For example, instead of using a constant symbol 0 one may use a predicate $0(x)$ (interpreted as $x=0$), and replace every predicate such as $P(0,y)$ with $\forall x\,(0(x)\rightarrow P(x,y))$. A function such as $f(x_1,x_2,...,x_n)$ will similarly be replaced by a predicate $F(x_1,x_2,...,x_n,y)$ interpreted as $y=f(x_1,x_2,...,x_n)$. This change requires adding additional axioms to the theory at hand, so that interpretations of the predicate symbols used have the correct semantics.

Restrictions such as these are useful as a technique to reduce the number of inference rules or axiom schemas in deductive systems, which leads to shorter proofs of metalogical results. The cost of the restrictions is that it becomes more difficult to express natural-language statements in the formal system at hand, because the logical connectives used in the natural language statements must be replaced by their (longer) definitions in terms of the restricted collection of logical connectives. Similarly, derivations in the limited systems may be longer than derivations in systems that include additional connectives. There is thus a trade-off between the ease of working within the formal system and the ease of proving results about the formal system.

It is also possible to restrict the arities of function symbols and predicate symbols, in sufficiently expressive theories. One can in principle dispense entirely with functions of arity greater than 2 and predicates of arity greater than 1 in theories that include a pairing function. This is a function of arity 2 that takes pairs of elements of the domain and returns an ordered pair containing them. It is also sufficient to have two predicate symbols of arity 2 that define projection functions from an ordered pair to its components. In either case it is necessary that the natural axioms for a pairing function and its projections are satisfied.

Many-sorted Logic

Ordinary first-order interpretations have a single domain of discourse over which all quantifiers range. Many-sorted first-order logic allows variables to have different sorts, which have different domains. This is also called typed first-order logic, and the sorts called types (as in data type), but it is not the same as first-order type theory. Many-sorted first-order logic is often used in the study of second-order arithmetic.

When there are only finitely many sorts in a theory, many-sorted first-order logic can be reduced to single-sorted first-order logic. One introduces into the single-sorted theory a unary predicate symbol for each sort in the many-sorted theory, and adds an axiom saying that these unary predicates partition the domain of discourse. For example, if there are two sorts, one adds predicate symbols $P_1(x)$ and $P_2(x)$ and the axiom

$$\forall x(P_1(x) \vee P_2(x)) \wedge \neg \exists x(P_1(x) \wedge P_2(x)).$$

Then the elements satisfying P_1 are thought of as elements of the first sort, and elements satisfying P_2 as elements of the second sort. One can quantify over each sort by using the corresponding predicate symbol to limit the range of quantification. For example, to say there is an element of the first sort satisfying formula $\varphi(x)$, one writes

$$\exists x(P_1(x) \wedge \phi(x)).$$

Additional Quantifiers

Additional quantifiers can be added to first-order logic.

- Sometimes it is useful to say that "$P(x)$ holds for exactly one x", which can be expressed as $\exists! x\, P(x)$. This notation, called uniqueness quantification, may be taken to abbreviate a formula such as $\exists x\, (P(x) \wedge \forall y\, (P(y) \rightarrow (x = y)))$.

- First-order logic with extra quantifiers has new quantifiers $Qx,...$, with meanings such as "there are many x such that ...". Also see branching quantifiers and the plural quantifiers of George Boolos and others.

- Bounded quantifiers are often used in the study of set theory or arithmetic.

Infinitary Logics

Infinitary logic allows infinitely long sentences. For example, one may allow a conjunction or disjunction of infinitely many formulas, or quantification over infinitely many variables. Infinitely long sentences arise in areas of mathematics including topology and model theory.

Infinitary logic generalizes first-order logic to allow formulas of infinite length. The most common way in which formulas can become infinite is through infinite conjunctions and disjunctions. However, it is also possible to admit generalized signatures in which function and relation symbols are allowed to have infinite arities, or in which quantifiers can bind infinitely many variables. Because an infinite formula cannot be represented by a finite string, it is necessary to choose some other representation of formulas; the usual representation in this context is a tree. Thus formulas are, essentially, identified with their parse trees, rather than with the strings being parsed.

The most commonly studied infinitary logics are denoted $L_{\alpha\beta}$, where α and β are each either cardinal numbers or the symbol ∞. In this notation, ordinary first-order logic is L. In the logic $L_{\infty\omega}$, arbitrary conjunctions or disjunctions are allowed when building formulas, and there is an unlimited supply of variables. More generally, the logic that permits conjunctions or disjunctions with less than κ constituents is known as $L_{\kappa\omega}$. For example, $L_{\omega_1\omega}$ permits countable conjunctions and disjunctions.

The set of free variables in a formula of $L_{\kappa\omega}$ can have any cardinality strictly less than κ, yet only finitely many of them can be in the scope of any quantifier when a formula appears as a subformula of another. In other infinitary logics, a subformula may be in the scope of infinitely many quantifiers. For example, in $L_{\kappa\infty}$, a single universal or existential quantifier may bind arbitrarily many variables simultaneously. Similarly, the logic $L_{\kappa\lambda}$ permits simultaneous quantification over fewer than λ variables, as well as conjunctions and disjunctions of size less than κ.

Non-classical and Modal Logics

- Intuitionistic first-order logic uses intuitionistic rather than classical propositional calculus; for example, $\neg\neg\varphi$ need not be equivalent to φ.

- First-order modal logic allows one to describe other possible worlds as well as this contingently true world which we inhabit. In some versions, the set of possible worlds varies depending on which possible world one inhabits. Modal logic has extra *modal operators* with meanings which can be characterized informally as, for example "it is necessary that φ" (true in all possible worlds) and "it is possible that φ" (true in some possible world). With standard first-order logic we have a single domain and each predicate is assigned one extension. With first-order modal logic we have a *domain function* that assigns each possible world its own domain, so that each predicate gets an extension only relative to these possible worlds. This allows us to model cases where, for example, Alex is a Philosopher, but might have been a Mathematician, and might not have existed at all. In the first possible world $P(a)$ is true, in the second $P(a)$ is false, and in the third possible world there is no a in the domain at all.

- First-order fuzzy logics are first-order extensions of propositional fuzzy logics rather than classical propositional calculus.

Fixpoint Logic

Fixpoint logic extends first-order logic by adding the closure under the least fixed points of positive operators.

Higher-order Logics

The characteristic feature of first-order logic is that individuals can be quantified, but not predicates. Thus

$$\exists a(\text{Phil}(a))$$

is a legal first-order formula, but

$$\exists \text{Phil}(\text{Phil}(a))$$

is not, in most formalizations of first-order logic. Second-order logic extends first-order logic by adding the latter type of quantification. Other higher-order logics allow quantification over even higher types than second-order logic permits. These higher types include relations between relations, functions from relations to relations between relations, and other higher-type objects. Thus the "first" in first-order logic describes the type of objects that can be quantified.

Unlike first-order logic, for which only one semantics is studied, there are several possible semantics for second-order logic. The most commonly employed semantics for second-order and higher-order logic is known as full semantics. The combination of additional quantifiers and the full semantics for these quantifiers makes higher-order logic stronger than first-order logic. In particular, the (semantic) logical consequence relation for second-order and higher-order logic is not semidecidable; there is no effective deduction system for second-order logic that is sound and complete under full semantics.

Second-order logic with full semantics is more expressive than first-order logic. For example, it is possible to create axiom systems in second-order logic that uniquely characterize the natural numbers and the real line. The cost of this expressiveness is that second-order and higher-order logics have fewer attractive metalogical properties than first-order logic. For example, the Löwenheim–Skolem theorem and compactness theorem of first-order logic become false when generalized to higher-order logics with full semantics.

Automated Theorem Proving and Formal Methods

Automated theorem proving refers to the development of computer programs that search and find derivations (formal proofs) of mathematical theorems. Finding derivations is a difficult task because the search space can be very large; an exhaustive search of every possible derivation is theoretically possible but computationally infeasible for many systems of interest in mathematics. Thus complicated heuristic functions are developed to attempt to find a derivation in less time than a blind search.

The related area of automated proof verification uses computer programs to check that human-created proofs are correct. Unlike complicated automated theorem provers, verification systems may be small enough that their correctness can be checked both by hand and through automated software verification. This validation of the proof verifier is needed to give confidence that any derivation labeled as "correct" is actually correct.

Some proof verifiers, such as Metamath, insist on having a complete derivation as input. Others, such as Mizar and Isabelle, take a well-formatted proof sketch (which may still be very long and detailed) and fill in the missing pieces by doing simple proof searches or applying known decision procedures: the resulting derivation is then verified by a small, core "kernel". Many such systems are primarily intended for interactive use by human mathematicians: these are known as proof assistants. They may also use formal logics that are stronger than first-order logic, such as type theory. Because a full derivation of any nontrivial result in a first-order deductive system will be extremely long for a human to write, results are often formalized as a series of lemmas, for which derivations can be constructed separately.

Automated theorem provers are also used to implement formal verification in computer science. In this setting, theorem provers are used to verify the correctness of programs and of hardware such as processors with respect to a formal specification. Because such analysis is time-consuming and thus expensive, it is usually reserved for projects in which a malfunction would have grave human or financial consequences.

Non-standard Models of Arithmetic

A nonstandard model of arithmetic is a proper elementary extension of the standard model $(N,0,1,+,\times,S)$ of first-order Peano arithmetic.

Existence of Non-standard Models

The non-standardness of N is expressible by axioms on an additional structure, forming an extended theory T'. One way is to add

- A constant symbol k with range N (which may need the axiom $k \in N$);

- The schema of axioms $n \neq k$ for each $n \in N$, making the number k non-standard.

If T is consistent then T' is also consistent (and cannot prove more), as any finite list of axioms of T' can be satisfied in any model of T (by interpreting k as a different number). Thus, T' also has a model, which is a model of T where N is non-standard.

This means that no consistent extension of arithmetic can ever force its model of arithmetic to be standard, i.e. require induction to apply to the full powerset of N, which it cannot ensure to exhaust by its classes. Candidate "standardness" predicates may be introduced but still cannot be forced to coincide with the true standardness (N_0), which may stay a meta-set beyond all classes.

Any model of arithmetic obtained as described in the proof of the completeness theorem from any consistent extension of arithmetic, will necessarily be non-standard, as it cannot even be elementarily equivalent to N according to the truth undefinability theorem. Skolem's paradox still holds in two ways:

- The meta-countable set P interpreting "$\wp(N)$", whose elements serve as subsets of N, still cannot exhaust $\wp(N)$ which is also meta-uncountable (on the meta level, bijections between sets N and N induce bijections between their powersets, which preserve uncountability);

- The image of P projected to $\wp(N)$ by taking preimages by the embedding from N to N, does not fill $\wp(N)$ either.

Non-standard models behave much like standard ones, as they satisfy all consequences of induction. Their precise range of diversity depends on the chosen theory of arithmetic.

Non-standard Models of Bare Arithmetic

The only "addition" definable from S, is the meta-operation adding any $n \in N$ to any $x \in N$ as $x + n = S^n(x)$, that is the meta-sequence of functions $(S^n)_{n \in N}$ each defined in the theory by a unary S-term n. The consequences of the induction schema in bare arithmetic can be summed up as

- The bijectivity of N as a $\{0,S\}$-algebra, which implies the bijectivity of S on the set of non-standard numbers;

- The schema of formulas $\forall n \in N^*, \forall x \in N, x + n \neq x$ which is a weak version of (F).

As the restriction of S to the meta-set $N\backslash N_0$ of non-standard numbers of any model is a permutation, $N\backslash N_0$ is a Z-set. The last formula obliges this Z-set to be free. Conversely, the disjoint union of N with any free Z-set, forms a model of bare arithmetic.

Applying both ways the definition of the order from the partial meta-operation of addition, gives two ordering results with non-standard numbers:

- Each Z-orbit of non-standard numbers has its own total order.

- Each non-standard number is greater than each standard number, thus may be called "infinitely big".

Models of arithmetic with order, are formed as models of bare arithmetic with any choice of a total order on the partition of non-standard numbers in Z-orbits. Thus, bare arithmetic cannot suffice to define the order. But arithmetic with order cannot suffice to define addition either, as its non-standard models may either admit no or many corresponding interpretations of addition.

Non-standard Models of Presburger Arithmetic

Such models satisfy all theorems of Presburger Arithmetic. In particular they have a well-defined total order, by which any non-standard number is greater than any standard number, and any non-empty class of numbers has a smallest element (which the meta-set of non-standard numbers hasn't).

There is also a meta-operation (sequence of functions) of multiplication of an $x \in N$ by a standard number $n \in N$: $n{\cdot}x = x+...+x$ (with n occurrences of x).

Beyond commutativity, associativity and the seen properties of the order, the last independent consequence of the axiom schema of induction constraining non-standard models, is the possibility of Euclidean division by any nonzero standard number (generalizing the results on parity):

$$\forall d \in N^*, \forall x \in N, \exists q \in N, \exists r < d, x = d{\cdot}q + r$$

thanks to $\forall d \in N^*, \forall q \in N, d{\cdot}(q+1) = d{\cdot}q+d$ which is a schema of theorems in Presburger arithmetic. Moreover this (q,r) is unique; $q = x{:}d \in N$ is called the *quotient* and r is called the *rest* of the division of an $x \in N$ by a $d \in N^*$, so that

$$\forall q \in N, q=x{:}d \Leftrightarrow d{\cdot}q \leq x < d{\cdot}(q+1) \Leftrightarrow \exists\, r < d, x = d{\cdot}q + r.$$

The concept of *model of Preburger arithmetic generated by a set* (of non-standard numbers, since standard ones have no generating effect), is defined as that of subalgebra generated by a subset for the "algebra" (not exactly an algebra, but...) with language completed, like in the proof of the completeness theorem, by additional operation symbols reflecting all existence properties deduced from the axiom schema of induction:

- The subtraction of a number by a lower number (or the absolute value of subtraction), which was implicit in the definition of the order and the proof that it is total;

- The sequence of functions $(x \mapsto x{:}d)$ of Euclidean division by each $d \in N^*$.

They can be conceived either in the abstract (by evaluating relations arbitrarily, like in the proof of the completeness theorem) or as a subset of a given model (interpreting expressions there). In particular, in any model of Preburger arithmetic generated by a single element (non-standard $k \in N$), the set of non-standard elements is the set of all values of expressions of the form:

$$(a \cdot k) : d + b$$

where $a, d \in N^*$ and $b \in Z$ (the cases where a, d are relative primes suffice). The predicate of divisibility of $x \in N$ by a $d \in N^*$, is defined as the case when the rest cancels: $d|x \Leftrightarrow (\exists q \in N, x = d \cdot q)$.

The possible shapes of these models with respect to k (classes of isomorphisms preserving k, described using it), are classified by the sequence $(r_d)_{d \in N^*}$ of rests of the division of k by all standard numbers d. The possible sequences are those which satisfy not only $r_d < d$ but also the compatibility formulas : $\forall n, d \in N^*, \exists h, r_{d \cdot n} = d \cdot h + r_d$ (where in fact $h < n$). The simplest one is where all r_d are 0, i.e. where k is divisible by every standard number (but the distinguishing property of this isomorphism class of models with 1 generator, "there exists a number divisible by every standard number", is inexpressible in Presburger arithmetic).

The non-standard models generated by 2 non-standard numbers k, k' can be split into the following classification depending on what may be intuitively described as the (standard) real number which k/k' is infinitely close to:

- k/k' may be infinitely large or infinitely small;

- It may be close to an irrational number;

- But the cases when it is close to a rational number a/b for standard numbers a, b, are reducible to other cases:

   ○ If the difference between $b \cdot k$ and $a \cdot k'$ is standard then the model is actually generated by only one element (any non-standard element is a generator).

   ○ If this difference is non-standard then, replacing one generator by this difference, reduces this model to the first case.

Structure of Countable Non-standard Models

The ultraproduct models are uncountable. One way to see this is to construct an injection of the infinite product of N into the ultraproduct. However, by the Löwenheim–Skolem theorem there must exist countable non-standard models of arithmetic. One way to define such a model is to use Henkin semantics.

Any countable non-standard model of arithmetic has order type $\omega + (\omega^* + \omega) \cdot \eta$, where ω is the order type of the standard natural numbers, ω^* is the dual order (an infinite decreasing sequence) and η is the order type of the rational numbers. In other words, a countable non-standard model begins with an infinite increasing sequence (the standard elements of the model). This is followed by a collection of "blocks," each of order type $\omega^* + \omega$, the order type of the integers. These blocks are in turn densely ordered with the order type of the rationals. The result follows fairly easily because it is easy to see that the non-standard numbers have to be dense and linearly ordered without endpoints, and the order type of the rationals is the only countable dense linear order without endpoints.

So, the order type of the countable non-standard models is known. However, the arithmetical operations are much more complicated.

It is easy to see that the arithmetical structure differs from $\omega + (\omega^* + \omega) \cdot \eta$. For instance if a non-standard (non-finite) element u is in the model, then so is $m \cdot u$ for any m, n in the initial segment N, yet u^2 is larger than $m \cdot u$ for any standard finite m.

Also you can define "square roots" such as the least v such that $v^2 > 2 \cdot u$. It is easy to see that these cannot be within a standard finite number of any rational multiple of u. By analogous methods to non-standard analysis you can also use PA to define close approximations to irrational multiples of a non-standard number u such as the least v with $v > \pi \cdot u$ (these can be defined in PA using non-standard finite rational approximations of π even though pi itself can't be). Once more, $v - (m/n) \cdot (u/n)$ has to be larger than any standard finite number for any standard finite m, n.

This shows that the arithmetical structure of a countable non-standard model is more complex than the structure of the rationals. There is more to it than that though.

Tennenbaum's theorem shows that for any countable non-standard model of Peano arithmetic there is no way to code the elements of the model as (standard) natural numbers such that either the addition or multiplication operation of the model is a computable on the codes. This result was first obtained by Stanley Tennenbaum in 1959.

Equational Logic

Formal languages are mathematical models of the natural (informal) languages of mathematics. In mathematical logic (i.e., meta-mathematics) one builds several classes of formal languages, of which first-order logic and equational logic are especially important. Languages of the first class are most often used to give complete mathematical definitions of mathematical theories, their axioms and their rules of proof. Languages of the second class are used most often in universal algebra, and in automatic theorem proving procedures.

An equational language L is a formal language whose alphabet consists of a countable set V of variables, a set Φ of function symbols and an equality symbol $=$. Moreover, a function $\rho : \Phi \to \{0,1,...\}$ is given and for each $f \in \Phi, \rho(f)$ denotes the number of argument places of f. If $\rho(f) = 0$, then f is called a constant.

One associates with L a class of algebras of type L, i.e. structures A of the form

$$\left(A, \tilde{f}\right)_{f \in \Phi},$$

where A is a non-empty set; if $\rho(f) > 0$, then $\tilde{f}$ is a function with $\rho(f)$ arguments running over A and with values in A; if $\rho(f) = 0$, then $\tilde{f} \in A$.

One defines the set T of terms of L to be the least set of finite sequences of letters of L such that T contains the one-term sequence consisting of a variable or a constant, and such that if $t1 ,...,t\rho (f) \in T$, then $f t1 ,...,t\rho (f) \in T$. If A is an algebra of type L and $t \in T$, then $\hat{t}$ denotes a composition of some of the functions and constants $\tilde{f}$, which is coded by t. A term of the form $f v1 ,...,v\rho (f)$, where $f \in \Phi$ and $v_i \in V$, is called atomic.

The only truth-valued expressions of L are equations, i.e., sequences of letters of the form

$$s = t,$$

where $s, t \in T$. One says that (a1) is true in A if and only if the objects $\tilde{s}$ and $\hat{t}$ are equal.

If E is a set of equations, then A is called a model of E if and only if all the equations of E are true in A. The class of all models of some set E is called a variety.

For any $s, t \in T$ and $v \in V$, one denotes by

$$s\binom{v}{t}$$

the term obtained from s by substituting all occurrences of v by t.

The rules of proof of L are the following:

i. $t = t$ is accepted for all $t \in T$;

ii. $s = t$ yields $t = s$;

iii. $r = s$ and $s = t$ yield $r = t$;

iv. $q = r$ and $s = t$ yield

$$q\binom{v}{s} = r\binom{v}{t}.$$

A set S of equations is called an equational theory if and only if S is closed under the rules i)–iv). Thus, if A is a model of E, then A is also a model of the least equational theory including E.

The above concepts and rules were introduced in 1935 by G. Birkhoff, and he proved the following fundamental theorems.

Birkhoff's completeness theorem: If S is an equational theory and $s = t$ is true in all the models of S, then $s = t$ belongs to S.

Birkhoff's characterization of varieties: A class C of algebras of type L is a variety if and only if it satisfies the following three conditions:

a) all subalgebras of the algebras of C are in C;

b) all homomorphic images of the algebras of C are in C;

c) for any subset K of C, the direct product of the algebras in K belongs to C.

These theorems are the roots of a very large literature.

In mathematical practice, as a rule one uses informal multi-sorted languages. Equational logic generalizes in a similar way. For example, a module over a ring is a two-sorted algebra with two universes, an Abelian group and a ring, and its language has two separate sorts of variables for the elements of those universes. Every model M of a first-order theory can be regarded as a two-sorted algebra whose universes are the universe of M and a two-element Boolean algebra, while treating the relations of M as Boolean-valued functions. Corresponding to that view there is a natural translation of any first-order language L^* into a two-sorted equational language L. Namely, the formulas of L^* are treated as Boolean-valued terms of L and the terms of L^* are treated as object-valued terms of L. The axioms and rules of first-order logic turn into the rules i)–iv) (adapted to the two-sorted language L) plus five axiom schemata:

1) $\left[(\varphi \rightarrow \psi) \rightarrow ((\psi \rightarrow x) \rightarrow (\varphi \rightarrow x))\right] = 1;$

2) $\left(\varphi \rightarrow (\neg\varphi \rightarrow \psi)\right) = 1;$

3) $\left((\neg\varphi \rightarrow \varphi) \rightarrow \varphi\right) = 1;$

4) $(1 \rightarrow \varphi) = \varphi;$

5) $\left(\varphi \rightarrow \varphi\left(\begin{matrix} x \\ \in x\varphi \end{matrix}\right)\right) = 1.$ Here $\varphi, \psi, ...$ run over formulas of L^*, the quantifiers of L^* are understood as abbreviations, viz.

$$\left(\exists x\,\varphi(x)\right) = \varphi\left(\begin{matrix} x \\ \in x\varphi \end{matrix}\right) \text{ and } \left(\forall x\,\varphi(x)\right) = \varphi\left(\begin{matrix} x \\ \in x(\neg\varphi) \end{matrix}\right),$$

where x is an object-valued variable, $\in x\varphi$ is an object-valued atomic term, called an $\in$-term, whose variables are the free variables of φ other than x, 1)–3) are equational versions of the Łukasiewicz axioms for propositional calculus, 4) yields the proper version of modus ponens, and 5) is the equational version of Hilbert's axiom about the $\in$-symbol (which he formulated in 1925).

In this way, the Gödel completeness theorem for first-order logic can be seen as a consequence of Birkhoff's completeness theorem stated. Moreover, equational logic corroborates a philosophical idea of H. Poincaré about the constructive and finitistic nature of mathematics. The same idea (in the context of set theory) was also expressed by D. Hilbert in 1904. Poincaré died in 1912, before the relevant mathematical concepts described above were developed by T. Skolem in 1920, Hilbert in 1925, and Birkhoff in 1935. Those concepts allow one to express this idea as follows. Quantifiers may suggest the actual existence of all objects of some infinite universes (a Platonic reality). But the above formalism shows that, at least in pure mathematics, they can be understood in a more concrete way, namely as abbreviations or blueprints for expressions involving certain ε-terms. And those ε-terms denote actually imagined objects or operations, thus they do not refer to nor imply the existence of any actually infinite universes. Hence the rules i)–iv) and the axiom schemata 1)–5) are constructive and finitistic in the sense of Poincaré and Hilbert.

Presently (1998), many researchers are trying to apply equational logic to obtain efficient automatic theorem proving procedures.

Second-order Logic

Second-order logic is an extension of first-order logic where, in addition to quantifiers such as "for every object (in the universe of discourse)," one has quantifiers such as "for every *property* of objects (in the universe of discourse)." This augmentation of the language increases its expressive strength, without adding new non-logical symbols, such as new predicate symbols. For classical extensional logic (as in this entry), properties can be identified with sets, so that second-order logic provides us with the quantifier "for every *set* of objects."

There are two approaches to the semantics of second-order logic. They differ on the interpretation of the phrase "for every set of objects." Does this have some fixed meaning to which we can refer, or do we need to consider the variety of meanings the phrase might have? In the first case (which will be called standard semantics), we are taking for granted certain mathematical concepts. In the second case (which will be called general semantics), much less is being taken for granted. In this case, to be considered *valid*, a sentence will need to be true under all the allowable meanings of the phrase "for every set of objects."

Syntax and Translation

In symbolic logic, the formula $(Px \rightarrow Px)$ will be true, no matter what object in the universe of discourse is assigned to the variable x. Which is to say nothing but that $\forall x(Px \rightarrow Px)$ will be true, no matter what subset of the universe of discourse is used to interpret the predicate symbol P. But is not that to say nothing but that the formula $\forall P \, \forall x(Px \rightarrow Px)$ is true, no matter what?

In first-order languages, there are some things we can say, and some that we cannot. Suppose, for example, that we want to express facts about the arithmetic of the natural numbers. That is, we want to express facts about the structure $(N; 0, S, <, +, \times)$ consisting of the set $N = \{0, 1, ...\}$ of natural numbers, together with the common arithmetical operations and relations. And we want to use a first-order language with quantifiers $\forall$ interpreted as "for every natural number" and $\exists$ interpreted as "for some natural number." Moreover, we include in the language a constant symbol 0 for the number zero, a one-place function symbol S for the successor operation (which applied to a natural number gives the next one), a two place predicate symbol < for the ordering relation < on N, and two-place function symbols + and × for addition and multiplication, respectively.

With this language, we can now symbolize many of the facts we know to be true about the natural numbers. We can form the sentence $\forall x(x < Sx)$ expressing the fact that each number is smaller than the next one, for example. But a difficulty arises if we want to express the "well-ordering property" that any non-empty set of natural numbers has a smallest member. If P is a new one-place predicate symbol, then

$$\exists x \, Px \rightarrow \exists x(Px \, \& \, \forall y(Py \rightarrow (y = x \lor x < y)))$$

expresses the idea that P is true of some smallest number, if it is true of any numbers at all. This formula is true in the structure $(N; 0, S, <, +, \times)$ when we interpret the predicate symbol P as being true of the numbers in some particular set—no matter what that set is—it says that the set has a least member, if it is non-empty. By now adding the quantifier ∀P

$$\forall P[\exists x\, Px \rightarrow \exists x(Px\, \&\forall y(Py \rightarrow (y = x \lor x < y)))]$$

we get a formalization of the well-ordering property.

The language of second-order logic extends the language of first-order logic by allowing quantification of predicate symbols and function symbols. As the foregoing example shows, in a second-order language for arithmetic, we can say that the natural numbers are well ordered. We know that the well-ordering property is not expressible by any first-order sentence, because the non-standard models of the (first-order) theory of (N; 0, S,< , +, ×) are never well ordered. So going to second-order logic is a genuine extension. That is, we can translate some natural-language sentences, such as "The relation< is a well-ordering," into the language of second-order logic that are not translatable into the language of first-order logic.

For another example, we can (using choice) say that the universe of discourse is infinite by saying that there is a transitive relation on the universe such that every element bears the relation to something, but not to itself:

$$\exists R[\forall x\, \forall y\, \forall z(Rxy\, \&\, Ryz \rightarrow Rxz)\, \&\, \forall x[\neg Rxx\, \&\exists y\, Rxy]$$

Here the second-order quantifier "∃R" expresses the existence of some binary relation on the universe. Because the only predicate symbol, R, is in the scope of this quantifier, the sentence has no predicate symbols open to interpretation. As is well known, no first-order sentence has for its models exactly the infinite structures.

In more detail, here is what is meant by a second-order language: One starts with a first-order language, and augments it by an unending supply of n-place predicate variables for each positive integer n, and an unending supply of n-place function variables for each positive integer n. (The function variables can be avoided, but that is another matter.) The formation rules for well-formed formulas are the obvious ones; in particular universal and existential quantification is allowed for any variable, be it an individual variable, a predicate variable, or a function variable.

To return to the example of natural numbers, we can express the Peano induction postulate by a second-order sentence:

$$\forall X[X0\, \&\, \forall y(Xy \rightarrow XSy) \rightarrow \forall y\, Xy]$$

This sentence expresses the idea that X is true of all natural numbers, if it is true of 0 and its truth at some number y guarantees its truth at the successor of y, no matter what set of numbers X might be true of.

For an example involving the set R of real numbers with its usual ordering, we can express the least-upper-bound property by a second-order sentence:

$$\forall X[\exists y\forall z(Xz \rightarrow z \leq y)\, \&\, \exists z\, Xz \rightarrow \exists y\forall y'(\forall z(Xz \rightarrow z \leq y') \leftrightarrow y \leq y')]$$

The foregoing examples are drawn from mathematical situations. There is also the intriguing possibility of natural-language sentences that seem to require second-order formulas for their formalization. George Boolos suggested the example, "There are some critics who admire only each other." This sentence asserts the existence of a set of individuals having a certain property; it does not entail, for example, that there are two critics who admire each other and admire no others.

Standard Semantics

Implicit is the concept of truth of a second-order sentence in a structure. Consider a structure M = (A, R, ...) consisting of a non-empty set A serving as the universe of discourse, and some relations and functions on A interpreting the non-logical symbols. Then we want to count a second-order sentence of the form $\forall P \, \varphi$ (where P is a k-place predicate variable) as being true in this structure if for every set Q of k-tuples of members of A, we have that φ is true in the structure when P is assigned the relation Q.

More formally, we need to define inductively what it means for a second-order formula φ to be satisfied in a structure M = (A, R, ...) under an assignment s of objects to the free variables in φ, which will be written $M \vDash \varphi[s]$. The definition proceeds exactly as in the first-order case, except for the additional clauses for the second-order quantifiers. For a k-place predicate variable P,

$$M \vDash \forall P \, \varphi[s] \text{ iff for every k-ary relation Q on A, we have } M \vDash \varphi[s']$$

where s' differs from s only in assigning the relation Q to the predicate variable P. (Here "iff" abbreviates "if and only if.") Similarly, for a k-place function variable F,

$$M \vDash \forall F \, \varphi[s] \text{ iff for every k-place function G on A, we have } M \vDash \varphi[s']$$

where s' differs from s only in assigning the function G to the function variable F. Observe that this definition refers, in the case of a 1-place predicate variable, to all subsets of A, that is, to the entire power set of A. It is this feature that accounts for the extraordinary semantical strength of second-order languages.

In the case of a second-order sentence σ (i.e., a formula with no free variable), the assignment s is no longer relevant, and we may speak unambiguously of the truth or falsity of σ in the structure M (that is, we can say that M is or is not a model of σ). In particular, the examples in §1 of translations from natural language into the language of second-order logic can now be seen to accomplish their intended purposes. The conjunction of the axioms for a linear ordering and the sentence,

$$\exists x \, Px \to \exists x(Px \,\&\, \forall y(Py \to (y = x \lor x < y)))$$

is true in a structure (A, <) iff the relation < well-orders the set A. The sentence

$$\exists R[\forall x \, \forall y \, \forall z(Rxy \,\&\, Ryz \to Rxz) \,\&\, \forall x(\neg Rxx \,\&\, \exists y \, Rxy)]$$

is true in a structure iff the universe of discourse is an infinite set. This example shows that the compactness theorem does not hold for second-order logic. If we call the above sentence $\lambda\infty$ and we let λn be the first-order sentence "there are at least n different things in the universe," then the set

$$\{\neg\lambda\infty, \quad \lambda 2, \quad \lambda 3, \quad \lambda 4, \quad ...\}$$

has no model, although each finite subset has a model.

The conjunction of the Peano postulates

$$\forall x(\neg 0 = Sx) \qquad \text{and} \qquad \forall x \forall y(Sx = Sy \to x = y)$$

and the Peano induction postulate

$$\forall X[X0 \ \& \ \forall y(Xy \to XSy) \to \forall y \ Xy]$$

is true in a structure (A, f, e) iff this structure is isomorphic to $(N, S, 0)$, the natural numbers with the successor operation S and distinguished element 0. The conjunction of these three sentences provides an example of a sentence that is categorical, that is, it has exactly one model, up to iso-morphism. By contrast, a first-order sentence can be categorical only if its one model is finite.

Similarly, the ordered field of real numbers, $(R, 0, 1, +, \times, <)$, can be characterized up to isomor-phism by the first-order axioms for an ordered field, together with the second-order sentence ex-pressing the least-upper-bound property. It is well known that any model of these sentences must be isomorphic to the ordered field of real numbers. This example shows that the Löwenheim–Skolem theorem does not hold for second-order logic.

The preceding examples show that two everyday mathematical structures, $(N, S, 0)$ and $(R, 0, 1, +, \times, <)$ are second-order characterizable. That is, each one has a single second-order axiom of which it is the only model, up to isomorphism. One might ask what other structures might be second-order characterizable. Of course, there can be only countably many such structures, up to isomorphism, because each one needs a sentence.

Next, suppose that in these examples, we existentially quantify all the non-logical symbols (i.e., all the predicate symbols and function symbols). Where $\pi(0, S,)$ is the conjunction of the three Peano postulates, the sentence $\exists x \ \exists F \ \pi(x, F)$ is a sentence in the second-order language of equality, that is, it has no non-logical symbols at all.

A structure for the language of equality consists simply of a non-empty universe of discourse; there are no relations or functions or distinguished elements. Such a structure is of course determined up to isomorphism simply by its cardinality. For a sentence σ in the language of equality, let its spec-trum be the class of cardinalities in which it is true. For example, the spectrum of a valid sentence is the class of all non-zero cardinal numbers. The spectrum of an unsatisfiable sentence is empty. The spectrum of $\neg\sigma$ is the complement (relative to the class of all non-zero cardinal numbers) of the spectrum of σ. Conjunction and disjunction of sentences yield intersection and union of spectra. A sentence in the language of equality is determined up to logical equivalence by its spectrum.

The sentence $\exists x \ \exists F \ \pi(x, F)$ we made from the Peano postulates is true in the countably infinite cardinality and no other one. Thus its spectrum is a singleton. Say that a cardinal number κ is second-order characterizable if there is a sentence of the second-order language of equality that is true in cardinality κ and only there. (The non-zero finite cardinals are all first-order characteriz-able.) We have seen that the countable infinite cardinal is second-order characterizable. Similarly, we can show that the power of the continuum is second-order characterizable. Where $\rho(0, 1, +, \times, <)$ is the second-order sentence that characterizes the ordered field of real numbers up to isomor-phism, the sentence,

$$\exists x \ \exists y \ \exists F \ \exists G \ \exists R \ \rho(x, y, F, G, R)$$

is a sentence in the second-order language of equality that is true in the power of the continuum and in no other cardinality.

One might ask what other cardinal numbers are second-order characterizable. See the 1974 paper by S. Garland for an exploration of this question. There can be only countably many such cardinals, of course, because each one takes a sentence.

It is not hard to see that the least uncountable cardinal is second-order characterizable. We can use a sentence saying that the universe is infinite but not countable, and that any uncountable subset is equinumerous to the entire universe. Thus we have sentences κ and λ in the second-order language of equality that characterize the least uncountable cardinal and the power of the continuum, respectively. The sentence $\kappa \leftrightarrow \lambda$ is a valid sentence if the continuum hypothesis is true, and only then. We can conclude that not all issues involving second-order logic are necessarily settled in ZFC.

The much-studied theory PA, first-order Peano arithmetic, is of course obtained by using, in place of the second-order induction postulate $\forall X[X0 \ \& \ \forall y(Xy \rightarrow XSy) \rightarrow \forall y \ Xy]$, the corresponding first-order schema

$$\varphi(0) \ \& \ \forall y(\varphi(y) \rightarrow \varphi(Sy)) \rightarrow \forall y \ \varphi(y)$$

where φ can be any suitable first-order formula. The effect of this schema is well known; it assures that any definable set that contains 0 and is closed under successor must contain everything.

Suppose that, by analogy, we start from our second-order axiomatization of the ordered field of real numbers, and replace the second-order least-upper-bound axiom by the corresponding schema. The result is an infinite set of first-order axioms, assuring that any definable set that is non-empty and bounded has a least upper bound. The models of this are called real-closed ordered fields. Interestingly, this concept was first formulated by algebraists, not by logicians.

One measurement of the strength of second-order logic is the complexity of its set of valid sentences. Let V^1 be the set of valid sentences of first-order logic and let V^2 be the set of valid sentences of second-order logic. More specifically, in first-order logic with only a single 2-place predicate symbol P, we know that the set $V^1(P)$ of valid sentences is a complete computably enumerable set (i.e., a complete recursively enumerable set). (Here we can assign Gödel numbers and view $V^1(P)$ as a set of natural numbers, or equivalently we can view it directly as a set of words over a finite alphabet.) And Tarski has pointed out that the set $V^1(=)$ of valid sentences in the first-order language of equality (with no non-logical symbols at all) is decidable.

For comparison, let $V^2(=)$ be the set of valid sentences in the second-order language of equality. What is the complexity of this set?

Let π be the conjunction of the Peano postulates and the recursion equations for addition and multiplication. Thus π is a second-order sentence in the language of arithmetic, with 0, S, +, and ×. The sentence π is categorical; its only model is $(N, 0, S, +, \times)$, up to isomorphism. Consequently, for a sentence σ in the language of arithmetic, σ is true in arithmetic iff the conditional $(\pi \rightarrow \sigma)$ is valid. This shows that $V^2(0, S, +, \times)$ cannot be arithmetical (i.e., cannot be first-order definable in arithmetic), lest truth in arithmetic be definable, in violation of Tarski's theorem. Now we can quantify away all the non-logical symbols; a sentence $\varphi(P)$ is valid iff the sentence $\forall P \ \varphi(P)$ is valid. The conclusion is that $V^2(=)$ is not arithmetical.

As interesting as that may be, it is merely the tip of the iceberg. To begin with, we can show that $V^2(=)$ is not analytical, that is, is not definable in arithmetic by a second-order formula. The proof of Tarski's theorem, showing that the set of true first-order sentences of arithmetic is not first-order definable in arithmetic, also shows that the set of true second-order sentences of arithmetic is not second-order definable in arithmetic. The rest of the argument is unchanged. And later we will see that even more is true.

In a very different direction, R. Fagin has shown a surprising connection between a topic in computational complexity and second-order definability over finite structures. For example, a finite graph can be regarded as a pair (V, E) consisting of a non-empty vertex set V and a symmetric edge relation E on V. The statement that the graph can be properly colored with three colors can be expressed by a second-order sentence: there exist subsets R, G, B that partition V in such a way that two vertices connected by an edge are never the same color. This sentence is Σ-1-1, that is, it has the form,

[existential second-order quantifiers] [first-order formula].

It is well known that being three-colorable is an NP property of a graph. That is, it is a property that is recognizable in polynomial time by a non-deterministic Turing machine. (There is a non-deterministic Turing machine M and a polynomial p such that whenever (V, E), suitably encoded, is given to M, then if (V, E) is three-colorable then some computation of M will accept the graph within p(n) steps, where n measures the size of (V, E), and if (V, E) is not three-colorable then no computation of M will ever accept the graph.)

Fagin showed that this is not an isolated example; every NP property of finite graphs is definable by a Σ-1-1 sentence of second-order logic. And conversely, any Σ-1-1 sentence defines an NP property. And in place of graphs, we can use directed graphs or other finite structures. Fagin's theorem states that a property of finite structures is an NP property if and only if it is definable by a Σ-1-1 second-order sentence.

General Semantics

A key feature of the "standard semantics" is that, for a one-place predicate variable X, the quantifier ∀X ranges over the entire power set of the universe of discourse. We have seen that this feature gives second-order languages a high degree of expressive strength.

But do we really want the quantifier ∀X to range over the actual power set? The predicativist will object that the power-set operation is not meaningful. And even the classical mathematician will admit that there are some obscure features of the power-set operation. The independence of the continuum hypothesis illustrates one such obscurity. If our goal is to study the foundations of mathematics, then it might be prudent not to take for granted that we already know all about power sets.

The concept of general semantics for second-order logic avoids any pretense that the power-set operation is a fixed well-understood resource. Instead, the range of the quantifier ∀X must be directly specified.

By a general pre-structure for a second-order language we mean a structure in the usual sense (a universe of discourse plus interpretations for the non-logical symbols) together with the additional sets:

- The n-place relation universe for each positive integer n. This must be a collection of n-ary relations on the universe of discourse. In particular, the 1-place relation universe must be some collection of subsets of the universe. Thus it is part (perhaps all, perhaps not) of the power set of the universe.

- The n-place function universe for each positive integer n. This must be a collection of n-place functions on the universe of discourse.

For a general pre-structure M, there is a natural way to define what it means for a second-order formula φ to be satisfied in a structure M under an assignment s of objects to the free variables in φ, which again will be written M ⊨ φ[s]. The second-order quantifiers are now defined to range over the corresponding universe. For a k-place predicate variable P,

M ⊨ ∀P φ[s] iff for every k-ary relation Q in the k-place relation universe, we have M ⊨ φ [s′]

where s′ differs from s only in assigning the relation Q to the predicate variable P. Similarly, for a k-place function variable F,

M ⊨ ∀F φ[s] iff for every k-place function G in the k-place function universe, we have M ⊨ φ [s′]

where s′ differs from s only in assigning the function G to the function variable F. In the case of a second-order sentence σ (i.e., a formula with no free variable), the assignment s is no longer relevant, and we may speak unambiguously of the truth or falsity of σ in the general pre-structure M (that is, we can say that M is or is not a model of σ).

But for second-order logic, we do not really want the 1-place relation universe to be an arbitrary collection of subsets of the universe. We might not know everything about the power-set operation, but we know some things about it. In effect, using general pre-structures amounts to treating a second-order language as a many-sorted first-order language.

There are some subsets of the universe that we know about, because we can define them. That is, suppose φ is a formula in which only the variable u occurs free. Then the set φ defines in M is the set consisting of all members a of M such that φ is satisfied in M when a is assigned to u. This idea can be extended. Suppose that φ has only the free variables u, v, w, x, Y, and Z (where Y is an m-place predicate variable and Z is an n-place function variable). Suppose that c and d are members of the (individual) universe |M| of M, that E is in M's m-place relation universe, and that F is in M's n-place function universe. Then the binary relation φ defines in M from the parameters c, d, E, and F is the set of pairs <a, b> of elements of |M| such that φ is satisfied in M when its variables u, v, w, x, Y, and Z are assigned a, b, c, d, E, and F, respectively. That is, it is the binary relation:

$$\{<a, b> \mid M ⊨ \varphi(u, v, w, x, Y, Z) [a, b, c, d, E, F]\}$$

Obviously, this concept can be generalized to the situation where a k-ary relation is defined from any particular number of parameters.

Then it is reasonable to restrict attention to general pre-structures that are closed under definability. Thus in the situation just described, it is reasonable to expect M's 2-ary relation universe to contain the binary relation that φ defines from parameters in the pre-structure. That is, we expect the sentence,

$$\forall w \, \forall x \, \forall Y \, \forall Z \, \exists R \, \forall u \, \forall v \, [Ruv \leftrightarrow \varphi(u, v, w, x, Y, Z)]$$

to be true in M. Call such sentences comprehension axioms.

By a general structure for a second-order language (also called a Henkin structure) is meant a general pre-structure in which all comprehension axioms (for all formulas) are true. (Here φ might contain quantifiers over predicate variables, so that even impredicative comprehension axioms are to be true. W e consider alternatives to "full comprehension.") Among the general structures are those in which the 1-place relation universe is the actual power set of the individual universe, and so forth. (Call such a general structure absolute.) But there can be others.

We obtain the general semantics (also called the Henkin semantics) for a second-order language by considering all general structures. That is, for a sentence σ to be valid in the general semantics, it must be true in all general structures. This is a stronger requirement than saying that σ is valid in the standard semantics. A sentence that is valid in the standard semantics is true in those general structures for which the 1-place relation universe is the full power set of the individual universe, and so forth. But such a sentence σ might turn out to be false in some general structure (that is, ¬σ might have a general model).

The main feature of the general semantics is a result of the "nothing but" type: Second-order logic with the general semantics is nothing but first-order logic (many-sorted) together with the comprehension axioms. Thus a sentence is valid in the general semantics iff it is logically implied (in first-order logic) by the set of comprehension axioms.

This reduction to first-order logic yields at once the following results:

- (Enumerability) In a second-order language with finitely many non-logical symbols, the set of sentences that are valid in the general semantics is computably enumerable. This holds because the set of comprehension axioms is a computable set (i.e., a recursive set).

- (Compactness) A set of sentences has a general model if every finite subset has a general model.

- (Löwenheim–Skolem) If a set of sentences has a general model, then it has a countable general model.

In each of these three cases, there is a sharp contrast to the situation of standard semantics considered. Moreover, a deductive calculus can be given for second-order logic (adapted from first-order logic and augmented by the comprehension axioms) that will be complete for the general semantics.

For comparison, axiomatic set theory (ZFC say) is a first-order theory; a model of set theory must supply a power-set operation. But the language of set theory has certain higher-order aspects, in that it permits us to speak of sets, sets of sets, and so forth.

In the extreme case of a structure M in which all relations are definable (e.g. a structure with a one-point universe), the general semantics will coincide with the standard semantics.

Higher-order Logic

There is no need to stop at second-order logic; one can keep going. We can add to the language "super-predicate" symbols, which take as arguments both individual symbols (either variables or constants) and predicate symbols. And then we can allow quantification over super-predicate symbols. And then we can keep going further.

(The reader is to be cautioned that there are in the literature two different ways of counting the order. According to one scheme, third-order logic allows super-predicate symbols to occur free, and fourth-order logic allows them to be quantified. According to the other scheme, third-order logic already allows quantification of super-predicate symbols.)

We reach the level of type theory after ω steps. And continuation into the transfinite is conceivable.

We have seen that, although the set V^1 of valid formulas of first-order logic is computably enumerable, the corresponding set V^2 for second-order logic (with the standard semantics) is vastly more complex. This phenomenon does not continue into the higher orders.

There is a sense in which the power-set operation is definable in second-order logic. Consider a language with a one-place predicate symbol I (for individuals), a one-place predicate symbol S (for sets), and a two-place predicate symbol E (for the membership relation). Then to express the idea that S is the power set of I we can use the conjunction (call it σ) of the following four sentences:

$\forall x(Ix + Sx)$ where "+" denotes exclusive disjunction
$\forall x\, \forall y\, (Exy \rightarrow Ix\,\&\, Sy)$
$\forall x\, \forall y\, (Sx\, \&\, Sy\, \&\, \forall t(Etx \leftrightarrow Ety) \rightarrow x = y)$ (extensionality)
$\forall X\, \exists y(Sy\, \&\, \forall t(It \rightarrow (Ety \leftrightarrow Xt)))$ (comprehension).

Clearly σ is true in any structure whose universe is the disjoint union of a set A and its power set P(A) and which assigns A to I, assigns P(A) to S, and assigns the membership relation $\in$ to E. Conversely, let M be any model of σ (in the standard semantics). Let f be a one-to-one function from M's interpretation of I onto a set A that is disjoint from its power set (there always is such a set). Extend f to all of M's universe by defining for each s in M's interpretation of S:

$f(s) = \{f(i) \mid M \vDash E\,[i, s]\}$

(that is, f(s) is the set of things in A that M thinks belong to s). Then f is an isomorphism from M to a structure whose universe is the disjoint union of a set A and its power set P(A) and which assigns A to I, assigns P(A) to S, and assigns the membership relation $\in$ to E. So roughly speaking, σ defines the power-set operation "to within isomorphism."

In a similar vein, we can define to within isomorphism the power set of $I \times I$, i.e., the set of binary relations on I. And so forth.

This second-order expressibility of the power-set operation permits the simulation of higher-order logic within second order. More specifically, we have the following result of Hintikka (1955): For each formula φ of higher-order logic (in a language with finitely many non-logical symbols), we can effectively find a sentence ψ of second-order logic (in the language of equality) such that φ is valid if and only if ψ is valid. The sentence ψ is constructed by first expanding the language by adding symbols for universes of various types (individuals, sets of individuals, ...) and for membership in these universes. Then φ's validity is equivalent to the validity of a second-order formula,

> The universes are correctly arranged $\rightarrow \varphi^*$

where φ^* is a suitable relativization of φ. Finally, we can prefix universal quantifiers to obtain a sentence ψ in the language of equality.

Thus the set of validities of seventeenth-order logic is computably reducible to $V^2(=)$, the set of second-order validities in the language of equality. (In fact, these two sets are computably isomorphic.) So in this aspect, the complexity of higher-order logic does not increase with the order. It follows that the set $V^2(=)$ has a high degree of complexity. We can extend our earlier observation that it is not definable in second-order arithmetic; it is not definable in arithmetic of higher order either. Montague in 1965 extended this into the transfinite. At the time, he was heard to say that the set $V^2(=)$ does not lie in any Kleene hierarchy, "past, present, or future."

The fact that we can express the power-set operation in second-order logic (and can iterate the procedure) gives second-order logic some large part of the expressiveness of set theory. Quine has claimed that second-order logic is not really logic, but rather set theory in disguise. And Robert Vaught has commented that studying second-order logic was like studying "the standard model of set theory."

The complexity of $V^2(=)$ does not change much if we impose limitations on the quantifier form of the sentences. The foregoing reduction of higher-order logic yields Π-1-2 sentences, so we can conclude that the set of valid Π-1-2 sentences in the language of equality is computably isomorphic to the full $V^2(=)$. That is about the best possible result. The set of Π-1-1 valid sentences is a complete computably enumerable set; once we drop the universal second-order quantifiers we are looking at first-order formulas. The set of Σ-1-1 valid sentences in the language of equality is a complete co-c.e. set, i.e., the complement of a computably enumerable set. (The sentence $\exists P \varphi(P)$ is true in every non-zero cardinality iff the elementary sentence $\varphi(P)$ has models of every finite size, a co-c.e. property. And to a Turing machine we can effectively assign an elementary sentence having models of every finite size iff the machine never halts.) The set of valid Σ-1-2 sentences in the language of equality is also computably isomorphic to the full $V^2(=)$, but this fact requires a different sort of proof.

Systems of Second-order Number Theory

The language of arithmetic (with 0, S, $<$, $+$, and $\times$) is important to the foundations of mathematics. As axioms, we can take the usual Peano postulates, including the second-order induction axiom. In the standard semantics, the only model of the Peano postulates, up to isomorphism, is the usual model of arithmetic. So the theory generated by these axioms (in the standard semantics) is simply the second-order theory of true arithmetic.

But suppose we consider instead the general semantics. The Peano postulates have general models that can differ from the usual model in either (or both) of two ways. We can employ the compactness theorem to construct non-standard general models of the Peano postulates containing infinitely large numbers. We can also find general models of the Peano postulates in which the universe of sets is less than the full power set of the individual universe (i.e., general models that are not absolute). Indeed, any countable general model must be of this kind.

In the context of general models, we add the additional axiom schema for choice:

$$\forall n\ \exists X\ \varphi(n,X) \to \exists Y\ \forall n\ \varphi(n, \{t \mid Ynt\})$$

prefixed by universal quantifiers as needed. (Here the formula that has been written as $\varphi(n, \{t \mid Ytn\})$ is obtained from $\varphi(n,X)$ by replacing each term Xu by the term Ynu.)

The Peano postulates are strong enough to provide us with pairing functions. Consequently, for a general model, its 1-place relation universe completely determines its k-place relation universe for each k.

The traditional terminology is to refer to second-order number theory as analysis. The name derives from the fact that it is possible to identify real numbers with sets of natural numbers. The second-order quantifiers over sets of natural numbers can then be viewed as quantifiers over the real numbers. The appropriateness of the name is open to question, but its usage is well established. Accordingly, by a model of analysis we will mean a general model of the Peano postulates with choice. As a general model, any model of analysis must of course satisfy all of the comprehension axioms; later we will consider weakening this requirement.

Let A2 be the theory generated by the Peano postulates with choice (in the general semantics). Then A2 is a complete computably enumerable subset of true second-order arithmetic. A2 contains every true Σ-0-1 sentence. It does not contain every true Π-0-1 sentence.

We can obtain a stronger theory by restricting attention to the models of analysis that differ from the usual model in only the second of the two ways described previously. That is, define an ω-model of analysis to be a model of analysis in which the individual universe is the actual set of natural numbers and the symbols 0 and S have their usual interpretations. (Consequently, the symbols <, +, and × have their usual interpretations.) The set universe of an ω-model of analysis must therefore be some part (possibly all) of the power set of the natural numbers, but it must be such that full comprehension is satisfied.

The motivation for considering ω-models can be stated as follow. We have a clear understanding—or so we like to think—of the set of natural numbers. But we do not have anything like the same understanding of the power set of the set of natural numbers. So it is reasonable to hold fixed the part we are sure of, but to leave open to interpretation the part we are not so sure about.

An ω-model of analysis is completely determined by its set universe. Because we have pairing functions that are first-order definable, the set universe will determine the binary relation universe, and so on. A Löwenheim–Skolem argument will show that there are ω-models of analysis with a countable set universe.

In any ω-model of analysis, the true first-order sentences are exactly the ones true in usual model. But ω-models can differ on second-order sentences. Let Aω be the theory of ω-models, that is, the set of sentences true in all ω-models of analysis. Then Aω extends A2. It is a complete Π-1-1 set. It contains every true Π-1-1 sentence; it does not contain every true Σ-1-1 sentence.

By the ω-rule we mean the infinitary rule of inference that infers $\forall x\ \varphi(x)$ from the premisses $\varphi(0)$, $\varphi(1)$, $\varphi(2)$, … . Suppose we add this rule to the usual logical apparatus, and see what is then deducible from the Peano postulates with choice. A "deduction" using the ω-rule will in general be infinite, but it must be well founded. It is not hard to see that any sentence deducible in this way is in Aω. Conversely, a completeness theorem holds: Every sentence in Aω has a deduction of the sort described.

In a 1961 paper, Andrzej Mostowski described a way of going one step further, toward a stronger theory still. Suppose we are willing to consider not just the order type ω as being sufficiently understood, but even the concept of being a well-ordering. Define a β-model of analysis to be an ω-model with the additional property that orderings in the model that appear to be well-orderings really are. That is, an ω-model M of analysis is a β-model if every ordering relation (on the natural numbers) in M with the property that non-empty sets in M always have least members, is in fact a well-ordering.

Then the set Aβ of sentences that are true in all β-models turns out to be a complete Π-1-2 set. It contains every true Π-1-2 sentence; it does not contain every true Σ-1-2 sentence. Clearly we have the inclusions,

$$A2 \subseteq A\omega \subseteq A\beta \subseteq \text{True Second-Order Arithmetic.}$$

For example, the number-theoretic part of a transitive ∈-model of ZF set theory will always be a β-model of analysis. But we can construct a β-model without strong assumptions. There is in fact a smallest β-model, and it can be constructed in a way reminiscent of Gödel's definition of the class L of constructible sets.

Finally, it should be mentioned that there are contexts in which theories of second-order arithmetic with less than full comprehension are suitable. For example, one can take the theory ACA given by the Peano postulates together with comprehension axioms for first-order formulas only. Such theories have been shown to be applicable to the study of "reverse mathematics."

Monadic Second Order Logic (MSO)

Monadic second-order logic (mso) is a logic with two types of quantifiers: one can quantify over elements, and one can quantify over sets of elements. One cannot, however, quantify over sets of pairs, or over sets triples, etc.

Our main interest here is to use mso to describe properties of undirected graphs. For example, suppose that we view a undirected graph as relational structure (i.e. a model as in logic), where the universe is the vertices and there is one binary relation $E(x,y)$ for the edges; this relation is symmetric. The following formula,

$$\forall x \forall y \, E(x,y)$$

says that the graph is a clique. The formula only quantifies over vertices, i.e. it uses only first-order quantification. Now we consider a formula which uses also set quantification. We adopt the convention that small variables $x, y, z, \ldots$ range over elements and big variables $X, Y, Z, \ldots$ range over sets of elements. The following formula says that input graph is not connected:

$$\underbrace{\exists X}_{\text{exists a set}} \; \underbrace{(\forall x \forall y \, x \in X \wedge E(x,y) \Rightarrow y \in X)}_{\text{is closed under neighbours}} \wedge \underbrace{(\exists x \, x \in X) \wedge (\exists x \, x \notin X)}_{\text{is neither empty nor full}}$$

The above formula illustrates all constructs in mso: one can quantify over elements, over sets of elements, one can test membership of elements in sets, and one can use the predicates available in the input model.

Here is another example: an mso formula which says that the input graph is three colourable:

$$\exists X_1 \exists X_2 \exists X_3 \quad \underbrace{\forall x \bigvee_i X_i}_{\text{every vertex is coloured}} \quad \wedge \quad \underbrace{\forall x \forall y \; E(x,y) \Rightarrow \bigvee_{i \neq j} X_i(x) \wedge X_j(x)}_{\text{every edge has endpoints with different colours}}$$

We say that a property of graphs (or more generally, structures over some vocabulary) is *mso definable* if there is a formula of mso which is true exactly in those graphs which have the property.

Model Checking of MSO

Model checking of mso formulas is the problem of checking if a given formula is true in a given structure (here, a graph). This problem is computationally hard: if we assume that the input is both the graph and the formula, then the problem is PSPACE complete (actually, it is PSPACE complete even when the graph is fixed, e.g. the one vertex graph, by encoding QBF in a straightforward way). If the formula is fixed and the graph is the only input, then the problem can be NP complete, as the example of 3-colorability shows. Since mso has built in negation, then we can also get coNP problems, e.g. non-3-colorability, and by using alternation of set quantifiers we can get problems that are complete for any level of the polynomial hierarchy. Summing up – in general, the model checking problem is hard.

Intuitionistic Logic

Intuitionistic logic encompasses the principles of logical reasoning which were used by L. E. J. Brouwer in developing his intuitionistic mathematics, beginning in . Because these principles also underly Russian recursive analysis and the constructive analysis of E. Bishop and his followers, intuitionistic logic may be considered the logical basis of constructive mathematics.

Philosophically, intuitionism differs from logicism by treating logic as a part of mathematics rather than as the foundation of mathematics; from finitism by allowing constructive reasoning about uncountable structures (e.g. monotone bar induction on the tree of potentially infinite sequences of natural numbers); and from platonism by viewing mathematical objects as mental constructs with no independent ideal existence. Hilbert's formalist program, to justify classical mathematics by reducing it to a formal system whose consistency should be established by finitistic (hence constructive) means, was the most powerful contemporary rival to Brouwer's developing intuitionism. In his 1912 essay Intuitionism and Formalism Brouwer correctly predicted that any attempt to prove the consistency of complete induction on the natural numbers would lead to a vicious circle.

Brouwer rejected formalism per se but admitted the potential usefulness of formulating general logical principles expressing intuitionistically correct constructions, such as modus ponens. Formal systems for intuitionistic propositional and predicate logic and arithmetic were developed by

Heyting , Gentzen and Kleene . Gödel proved the equiconsistency of intuitionistic and classical theories. Kripke provided a semantics with respect to which intuitionistic logic is correct and complete.

Rejection of Tertium Non Datur

Intuitionistic logic can be succinctly described as classical logic without the Aristotelian law of excluded middle (LEM): $(A \lor \neg A)$ or the classical law of double negation elimination $(\neg \neg A \to A)$, but with the law of contradiction $(A \to B) \to ((A \to \neg B) \to \neg A)$ and ex falso quodlibet: $(\neg A \to (A \to B))$. Brouwer observed that LEM was abstracted from finite situations, then extended without justification to statements about infinite collections. For example, let x, y range over the natural numbers 0, 1, 2, ... and B(x) abbreviate the property expressed by the following claim in which the variable x is free: there is a y greater than x such that both y and $y+2$ are prime numbers, i.e.,

$$\exists y(y>x \ \& \ Prime(y) \ \& \ Prime(y+2))$$

Then we have no general method for deciding whether B(x) is true or false for arbitrary x, so $\forall x(B(x) \lor \neg B(x))$ cannot be asserted in the present state of our knowledge. And if A abbreviates the statement $\forall x B(x)$, then $(A \lor \neg A)$ cannot be asserted because neither A nor $(\neg A)$ has yet been proved.

One may object that these examples depend on the fact that the Twin Primes Conjecture has not yet been settled. A number of Brouwer's original "counterexamples" depended on problems (such as Fermat's Last Theorem) which have since been solved. But to Brouwer the general LEM was equivalent to the a priori assumption that every mathematical problem has a solution — an assumption he rejected, anticipating Gödel's incompleteness theorem by a quarter of a century.

The rejection of LEM has far-reaching consequences. On the one hand,

- Intuitionistically, Reductio ad absurdum only proves negative statements, since $\neg \neg A \to A$ does not hold in general. (If it did, LEM would follow by modus ponens from the intuitionistically provable $\neg \neg (A \lor \neg A)$.)

- Not every propositional formula has an intuitionistically equivalent disjunctive or conjunctive normal form.

- Not every predicate formula has an intuitionistically equivalent prenex form.

- While $\forall x \ \neg \ \neg \ (A(x) \lor \neg A(x))$ is a theorem of intuitionistic predicate logic, $\neg \ \neg \ \forall x(A(x) \lor \neg A(x))$ is not.

- Pure intuitionistic logic is axiomatically incomplete. Infinitely many intermediate axiomatic extensions of intuitionistic propositional and predicate logic are contained in classical logic.

On the other hand,

- Every intuitionistic proof of a closed statement of the form $A \lor B$ can be effectively transformed into an intuitionistic proof of A or an intuitionistic proof of B, and similarly for closed existential statements.

- Classical logic is finitistically interpretable in the negative fragment of intuitionistic logic.

- Arithmetical formulas have relatively simple intuitionistic normal forms.

- Intuitionistic arithmetic can consistently be extended by axioms (such as Church's Thesis) which contradict classical arithmetic, enabling the formal study of recursive mathematics.

- Brouwer's controversial intuitionistic analysis, which conflicts with LEM, can be formalized and shown consistent relative to a classically and intuitionistically correct subtheory.

Intuitionistic First-Order Predicate Logic

Formalized intuitionistic logic is naturally motivated by the informal Brouwer-Heyting-Kolmogorov explanation of intuitionistic truth, the entry on intuitionism in the philosophy of mathematics and discussed extensively in the entry on the development of intuitionistic logic. The constructive independence of the logical operations &, ∨, →, ¬, ∀, ∃ contrasts with the classical situation, where e.g., (A ∨ B) is equivalent to ¬ (¬A & ¬B), and ∃xA(x) is equivalent to ¬ ∀x ¬A(x). From the B-H-K viewpoint, a sentence of the form (A ∨ B) asserts that either a proof of A, or a proof of B, has been constructed; while ¬ (¬A& ¬B) asserts that an algorithm has been constructed which would effectively convert any pair of constructions proving ¬A and ¬B respectively, into a proof of a known contradiction.

The Formal Systems H–IPC and H–IQC

Following is a Hilbert-style formalism H–IQC, from Kleene, for intuitionistic first-order predicate logic IQC. The language L of H–IQC has predicate letters P, Q(.),... of all arities and individual variables a, b, c,... (with or without subscripts 1,2,...), as well as symbols &, ∨, →, ¬, ∀, ∃ for the logical connectives and quantifiers, and parentheses (,). The prime formulas of L are expressions such as P, Q(a), R(a, b, a) where P, Q(.), R(...) are 0-ary, 1-ary and 3-ary predicate letters respectively; that is, the result of filling each blank in a predicate letter by an individual variable symbol is a prime formula. The (well-formed) formulas of L are defined inductively as follows.

- Each prime formula is a formula.

- If A and B are formulas, so are (A & B), (A ∨ B), (A → B) and ¬A.

- If A is a formula and x is a variable, then ∀xA and ∃xA are formulas.

In general, we use A, B, C as metavariables for well-formed formulas and x, y, z as metavariables for individual variables. Anticipating applications (for example to intuitionistic arithmetic) we use s, t as metavariables for terms; in the case of pure predicate logic, terms are simply individual variables. An occurrence of a variable x in a formula A is bound if it is within the scope of a quantifier ∀x or ∃x, otherwise free. Intuitionistically as classically, "(A ↔ B)" abbreviates "((A → B) & (B → A))," and parentheses are omitted when this causes no confusion.

There are three rules of inference:

- Modus Ponens: From A and (A → B), conclude B.

- ∀-Introduction: From (C → A(x)), where x is a variable which does not occur free in C, conclude (C → ∀xA(x)).

- ∃-Elimination: From $(A(x) \to C)$, where x is a variable which does not occur free in C, conclude $(\exists x A(x) \to C)$.

The axioms are all formulas of the following forms, where in the last two schemas the subformula A(t) is the result of substituting an occurrence of the term t for every free occurrence of x in A(x), and no variable free in t becomes bound in A(t) as a result of the substitution.

- $A \to (B \to A)$.

- $(A \to B) \to ((A \to (B \to C)) \to (A \to C))$.

- $A \to (B \to A \,\&\, B)$.

- $A \,\&\, B \to A$.

- $A \,\&\, B \to B$.

- $A \to A \vee B$.

- $B \to A \vee B$.

- $(A \to C) \to ((B \to C) \to (A \vee B \to C))$.

- $(A \to B) \to ((A \to \neg B) \to \neg A)$.

- $\neg A \to (A \to B)$.

- $\forall x A(x) \to A(t)$.

- $A(t) \to \exists x A(x)$.

A proof is any finite sequence of formulas, each of which is an axiom or an immediate consequence, by a rule of inference, of (one or two) preceding formulas of the sequence. Any proof is said to prove its last formula, which is called a theorem or provable formula of first-order intuitionistic predicate logic. A derivation of a formula E from a collection F of assumptions is any sequence of formulas, each of which belongs to F or is an axiom or an immediate consequence, by a rule of inference, of preceding formulas of the sequence, such that E is the last formula of the sequence. If such a derivation exists, we say E is derivable from F.

Intuitionistic propositional logic H–IPC is the subtheory of H–IQC which results when the language is restricted to formulas built from proposition letters P, Q, R,... using the propositional connectives &, ∨, → and ¬, and only the propositional postulates are used. Thus the last two rules of inference and the last two axiom schemas are absent from the propositional theory.

If, in the given list of axiom schemas for intuitionistic propositional or first-order predicate logic, the law expressing ex falso sequitur quodlibet,

$$\neg A \to (A \to B)$$

is replaced by the classical law of double negation elimination:

$$\neg \neg A \to A$$

(or,equivalently, if the intuitionistic law of negation introduction;

$$(A \rightarrow B) \rightarrow ((A \rightarrow \neg B) \rightarrow \neg A)$$

is replaced by LEM), a formal system H–CPC for classical propositional logic CPC or classical predicate logic or H–CQC results. Since the law of contradiction is a classical theorem, intuitionistic logic is contained in classical logic. In a sense, classical logic is also contained in intuitionistic logic.

It is important to note that while LEM and the law of double negation are equivalent as schemas over H–IPC, the implication

$$(\neg \neg A \rightarrow A) \rightarrow (A \vee \neg A)$$

is not a theorem schema of H–IPC. For theories T based on intuitionistic logic, if E is an arbitrary formula of L(T) then by definition:

- E is decidable in T if and only if T proves $(E \vee \neg E)$.

- E is stable in T if and only if T proves $(\neg \neg E \rightarrow E)$.

- E is testable in T if and only if T proves $(\neg E \vee \neg \neg E)$.

Decidability implies stability, but not conversely. The conjunction of stability and testability is equivalent to decidability. By Brouwer's first published logical theorem $\neg \neg \neg A \rightarrow \neg A$, every formula of the form $\neg A$ is stable; but in H–IPC and H–IQC prime formulas and their negations are undecidable.

Alternative Formalisms and the Deduction Theorem

The Hilbert-style system H–IQC is useful for metamathematical investigations of intuitionistic logic, but its forced linearization of deductions and its preference for axioms over rules make it an awkward instrument for establishing derivability. A natural deduction system N–IQC for intuitionistic predicate logic results from the deductive system D, by omitting the symbol and rules for identity, and replacing the classical rule (DNE) of double negation elimination by the intuitionistic negation elimination rule,

(INE) If F entails A and F entails ¬A, then F entails B.

While identity can of course be added to intuitionistic logic, for applications (e.g., to arithmetic) the equality symbol is generally treated as a distinguished predicate constant satisfying nonlogical axioms (e.g., the primitive recursive definitions of addition and multiplication) in addition to reflexivity, symmetry and transitivity. Identity is decidable, intuitionistically as well as classically, but intuitionistic extensional equality is not always decidable.

The keys to proving that H–IQC is equivalent to N–IQC I are modus ponens and its converse, the:

Deduction Theorem

If B is derivable from A and possibly other formulas F, with all variables free in A held constant in the derivation (that is, without using the second or third rule of inference on any variable x occurring free in A, unless the assumption A does not occur in the derivation before the inference in question), then $(A \rightarrow B)$ is derivable from F.

This fundamental result, roughly expressing the rule $(\rightarrow I)$ of I, can be proved for H–IQC by induction on the definition of a derivation. The other rules of I hold for H–IQC essentially by modus ponens, which corresponds to $(\rightarrow E)$ in N–IQC. To illustrate the usefulness of the Deduction Theorem, consider the (apparently trivial) theorem schema $(A \rightarrow A)$ of IPC. A correct proof in H takes five lines:

1. $A \rightarrow (A \rightarrow A)$

2. $(A \rightarrow (A \rightarrow A)) \rightarrow ((A \rightarrow ((A \rightarrow A) \rightarrow A)) \rightarrow (A \rightarrow A))$

3. $(A \rightarrow ((A \rightarrow A) \rightarrow A)) \rightarrow (A \rightarrow A)$

4. $A \rightarrow ((A \rightarrow A) \rightarrow A)$

5. $A \rightarrow A$

where 1, 2 and 4 are axioms and 3, 5 come from earlier lines by modus ponens. However, A is derivable from A (as assumption) in one obvious step, so the Deduction Theorem allows us to conclude that a proof of $(A \rightarrow A)$ exists. (In fact, the formal proof of $(A \rightarrow A)$ just presented is part of the constructive proof of the Deduction Theorem.)

It is important to note that, in the definition of a derivation from assumptions in H, the assumption formulas are treated as if all their free variables were universally quantified, so that $\forall x\, A(x)$ is derivable from the hypothesis $A(x)$. However, the variable x will be varied (not held constant) in that derivation, by use of the rule of $\forall$-introduction; and so the Deduction Theorem cannot be used to conclude (falsely) that $A(x) \rightarrow \forall x\, A(x)$ (and hence, by $\exists$-elimination, $\exists x\, A(x) \rightarrow \forall x\, A(x)$) are provable in H. As an example of a correct use of the Deduction Theorem for predicate logic, consider the implication $\exists x\, A(x) \rightarrow \neg\forall x\neg A(x)$. To show this is provable in IQC, we first derive $\neg\forall x\neg A(x)$ from $A(x)$ with all free variables held constant:

1. $\forall x\neg A(x) \rightarrow \neg A(x)$

2. $A(x) \rightarrow (\forall x\neg A(x) \rightarrow A(x))$

3. $A(x)$ (assumption)

4. $\forall x\neg A(x) \rightarrow A(x)$

5. $(\forall x\neg A(x) \rightarrow A(x)) \rightarrow ((\forall x\neg A(x) \rightarrow \neg A(x)) \rightarrow \neg\forall x\neg A(x))$

6. $(\forall x\neg A(x) \rightarrow \neg A(x)) \rightarrow \neg\forall x\neg A(x)$

7. $\neg\forall x\neg A(x)$

Here 1, 2 and 5 are axioms; 4 comes from 2 and 3 by modus ponens; and 6 and 7 come from earlier lines by modus ponens; so no variables have been varied. The Deduction Theorem tells us there is a proof P in IQC of $(A(x) \rightarrow \neg\forall x\neg A(x))$, and one application of $\exists$-elimination converts P into a proof of $\exists x\, A(x) \rightarrow \neg\forall x\neg A(x)$. The converse is not provable in IQC.

Intuitionistic Number Theory (Heyting Arithmetic)

Intuitionistic (Heyting) arithmetic HA and classical (Peano) arithmetic PA share the same first-order language and the same non-logical axioms; only the logic is different. In addition to the logical connectives, quantifiers and parentheses and the individual variables a, b, c, ... (with metavariables x, y, z as usual), the language L(HA) of arithmetic has a binary predicate symbol =, individual constant 0, unary function constant S, and finitely or countably infinitely many additional constants for primitive recursive functions including addition and multiplication; the precise choice is a matter of taste and convenience. Terms are built from variables and 0 using the function constants; in particular, each natural number n is expressed in the language by the numeral n obtained by applying S n times to 0 (e.g., S(S(0)) is the numeral for 2). Prime formulas are of the form (s = t) where s, t are terms, and compound formulas are obtained from these as usual.

The logical axioms and rules of HA are those of first-order intuitionistic predicate logic IQC. The nonlogical axioms include the reflexive, symmetric and transitive properties of =, primitive recursive defining equations for each function constant, the axioms characterizing 0 as the least natural number and S as a one-to-one function:

- $\forall x \neg (S(x) = 0)$,
- $\forall x \forall y (S(x) = S(y) \to x = y)$,

the extensional equality axiom for S:

- $\forall x \forall y (x = y \to S(x) = S(y))$,

and the (universal closure of the) schema of mathematical induction, for arbitrary formulas A(x):

- $A(0) \,\&\, \forall x (A(x) \to A(S(x))) \to \forall x\, A(x)$.

Extensional equality axioms for all the other function constants are derivable by mathematical induction from the equality axiom for S and the primitive recursive function axioms.

The natural order relation x < y can be defined in HA by $\exists z(S(z) + x = y)$, or by a quantifier-free formula if the symbol and defining axioms for cutoff subtraction are present in the formalism. HA proves the comparative law

$$\forall x\, \forall y\, (x < y \lor x = y \lor y < x)$$

and an intuitionistic form of the least number principle, for arbitrary formulas A(x):

$$\forall x [\forall y\, (y < x \to A(y) \lor \neg A(y)) \to \exists y\, (y < x \,\&\, A(y) \,\&\, \forall z(z < y \to \neg A(z))) \lor \forall y(y < x \to \neg A(y))].$$

The hypothesis is needed because not all arithmetical formulas are decidable in HA. However, $\forall x \forall y (x = y \lor \neg(x = y))$ can be proved directly by mathematical induction, and so

- Prime formulas (and hence all quantifier-free formulas) are decidable and stable in HA.

If A(x) is decidable in HA, then by induction on x so are $\forall y\, (y < x \to A(y))$ and $\exists y\, (y < x \,\&\, A(y))$. Hence

- Formulas in which all quantifiers are bounded are decidable and stable in HA.

The collection Δ_0 of arithmetical formulas in which all quantifiers are bounded is the lowest level of a classical arithmetical hierarchy based on the pattern of alternations of quantifiers in a prenex formula. In HA not every formula has a prenex form, but Burr discovered a simple intuitionistic arithmetical hierarchy corresponding level by level to the classical. For the purposes of the next two definitions only, $\forall x$ denotes a block of finitely many universal number quantifiers, and similarly $\exists x$ denotes a block of finitely many existential number quantifiers. With these conventions, Burr's classes Φ_n and Ψ_n are defined by

- $\Phi_0 = \Psi_0 = \Delta_0$,

- Φ_1 is the class of all formulas of the form $\forall x\, A(x)$ where $A(x)$ is in Ψ_0. For $n \geq 2$, Φ_n is the class of all formulas of the form $\forall x\, [A(x) \rightarrow \exists y\, B(x,y)]$ where $A(x)$ is in Φ_{n-1} and $B(x,y)$ is in Φ_{n-2},

- Ψ_1 is the class of all formulas of the form $\exists x\, A(x)$ where $A(x)$ is in Φ_0. For $n \geq 2$, Ψ_n is the class of all formulas of the form $A \rightarrow B$ where A is in Φ_n and B is in Φ_{n-1}.

The corresponding classical prenex classes are defined more simply:

- $\Pi_0 = \Sigma_0 = \Delta_0$,

- Π_{n+1} is the class of all formulas of the form $\forall x\, A(x)$ where $A(x)$ is in Σ_n,

- Σ_{n+1} is the class of all formulas of the form $\exists x\, A(x)$ where $A(x)$ is in Π_n.

Peano arithmetic PA comes from Heyting arithmetic HA by adding LEM or $(\neg\neg A \rightarrow A)$ to the list of logical axioms, i.e., by using classical instead of intuitionistic logic. The following results hold even in the fragments of HA and PA with the induction schema restricted to Δ_0 formulas.

Burr's Theorem:

- Every arithmetical formula is provably equivalent in HA to a formula in one of the classes Φ_n.

- Every formula in Φ_n is provably equivalent in PA to a formula in Π_n, and conversely.

- Every formula in Ψ_n is provably equivalent in PA to a formula in Σ_n, and conversely.

HA and PA are proof-theoretically equivalent. Each is capable of (numeralwise) expressing its own proof predicate. By Gödel's famous Incompleteness Theorem, if HA is consistent then neither HA nor PA can prove its own consistency.

Basic Proof Theory

Translating Classical into Intuitionistic Logic

A fundamental fact about intuitionistic logic is that it has the same consistency strength as classical logic. For propositional logic this was first proved by Glivenko.

Glivenko's Theorem: An arbitrary propositional formula A is classically provable, if and only if $\neg\neg A$ is intuitionistically provable.

Glivenko's Theorem does not extend to predicate logic, although an arbitrary predicate formula A is classically provable if and only if $\neg\neg$A is provable in intuitionistic predicate logic plus the "double negation shift" schema

(DNS) $\forall x\neg\neg B(x) \rightarrow \neg\neg\forall x\, B(x)$.

The more sophisticated negative translation of classical into intuitionistic theories, due independently to Gödel and Gentzen, associates with each formula A of the language L another formula g(A) (with no $\vee$ or $\exists$), such that

(I)   Classical predicate logic proves $A \leftrightarrow g(A)$.

(II)   Intuitionistic predicate logic proves $g(A) \leftrightarrow \neg\neg\, g(A)$.

(III)   If classical predicate logic proves A, then intuitionistic predicate logic proves g(A).

The proofs are straightforward from the following inductive definition of g(A) (using Gentzen's direct translation of implication, rather than Gödel's in terms of $\neg$ and &):

- g(P) is $\neg\neg$ P, if P is prime.
- g(A & B) is (g(A)& g(B)).
- g(A $\vee$ B) is $\neg$ ($\neg$g(A) & $\neg$g(B)).
- g(A $\rightarrow$ B) is (g(A) $\rightarrow$ g(B)).
- g($\neg$A) is $\neg$ g(A).
- g($\forall$xA(x)) is $\forall$x g(A(x)).
- g($\exists$xA(x)) is $\neg\forall$x$\neg$g(A(x)).

For each formula A, g(A) is provable intuitionistically if and only if A is provable classically. In particular, if (B & $\neg$B) were classically provable for some formula B, then (g(B)& $\neg$g(B)) (which is g(B& $\neg$B)) would in turn be provable intuitionistically. Hence

(IV) Classical and intuitionistic predicate logic are equiconsistent.

The negative translation of classical into intuitionistic number theory is even simpler, since prime formulas of intuitionistic arithmetic are stable. Thus g(s=t) can be taken to be (s=t), and the other clauses are unchanged. The negative translation of any instance of mathematical induction is another instance of mathematical induction, and the other nonlogical axioms of arithmetic are their own negative translations, so

(I), (II), (III) and (IV) hold also for number theory.

Gödel [1933e] interpreted these results as showing that intuitionistic logic and arithmetic are richer than classical logic and arithmetic, because the intuitionistic theory distinguishes formulas which are classically equivalent, and has the same consistency strength as the classical theory.

Direct attempts to extend the negative interpretation to analysis fail because the negative translation of the countable axiom of choice is not a theorem of intuitionistic analysis. However, it is consistent with intuitionistic analysis, including Brouwer's controversial continuity principle, by the functional version of Kleene's recursive realizability.

Admissible Rules of Intuitionistic Logic and Arithmetic

Gödel observed that intuitionistic propositional logic has the disjunction property:

(DP) If $(A \lor B)$ is a theorem, then A is a theorem or B is a theorem.

Gentzen established the disjunction property for closed formulas of intuitionistic predicate logic. From this it follows that if intuitionistic logic is consistent, then $(P \lor \neg P)$ is not a theorem if P is prime. Kleene [1945, 1952] proved that intuitionistic first-order number theory also has the related existence property:

(ED) If $\exists x A(x)$ is a closed theorem, then for some closed term t, $A(t)$ is a theorem.

The disjunction and existence properties are special cases of a general phenomenon peculiar to nonclassical theories. The admissible rules of a theory are the rules under which the theory is closed. For example, Harrop observed that the rule

If $(\neg A \to (B \lor C))$ is a theorem, so is $(\neg A \to B) \lor (\neg A \to C)$

is admissible for intuitionistic propositional logic IPC because if A, B and C are any formulas such that $(\neg A \to (B \lor C))$ is provable in IPC, then also $(\neg A \to B) \lor (\neg A \to C)$ is provable in IPC. Harrop's rule is not derivable in IPC because $(\neg A \to (B \lor C)) \to (\neg A \to B) \lor (\neg A \to C)$ is not intuitionistically provable. Another important example of an admissible nonderivable rule of IPC is Mints' rule:

If $((A \to B) \to A \lor C)$ is a theorem, so is $((A \to B) \to A) \lor ((A \to B) \to C)$.

The two-valued truth table interpretation of classical propositional logic CPC gives rise to a simple proof that every admissible rule of CPC is derivable: otherwise, some assignment to A, B, etc. would make the hypothesis true and the conclusion false, and by substituting e.g. $(P \to P)$ for the letters assigned "true" and $(P \, \& \, \neg P)$ for those assigned "false" one would have a provable hypothesis and unprovable conclusion. The fact that the intuitionistic situation is more interesting leads to many natural questions, some of which have recently been answered.

By generalizing Mints' Rule, Visser and de Jongh identified a recursively enumerable sequence of successively stronger admissible rules ("Visser's rules") which, they conjectured, formed a basis for the admissible rules of IPC in the sense that every admissible rule is derivable from the disjunction property and one of the rules of the sequence. Building on work of Ghilardi , Iemhoff succeeded in proving their conjecture. Rybakov proved that the collection of all admissible rules of IPC is decidable but has no finite basis. Visser showed that his rules are also the admissible propositional rules of HA, and of HA extended by Markov's Principle MP. More recently, Jerabek found a different basis for the admissible rules of IPC with the property that no rule in the basis derives another.

Much less is known about the admissible rules of intuitionistic predicate logic. Pure IQC, without individual or predicate constants, has the following remarkable admissible rule for A(x) with no variables free but x:

If ∃x A(x) is a theorem, so is ∀x A(x).

Not every admissible predicate rule of IQC is admissible for all formal systems based on IQC; for example, HA evidently violates the rule just stated. Visser proved in that the property of being an admissible predicate rule of HA is Π_2 complete, and in that HA + MP has the same predicate admissible rules as HA. Plisko proved that the predicate logic of HA + MP (the set of sentences in the language of IQC all of whose uniform substitution instances in the language of arithmetic are provable in HA + MP) is Π_2 complete; Visser extended this result to some constructively interesting consistent extensions of HA which are not contained in PA.

While they have not been completely classified, the admissible rules of intuitionistic predicate logic are known to include Markov's Rule for decidable predicates:

If ∀x(A(x) ∨ ¬A(x)) & ¬∀x¬A(x) is a theorem, so is ∃x A(x)

and the following Independence-of-Premise Rule (where y is assumed not to occur free in A(x)):

If ∀x(A(x) ∨ ¬A(x)) & (∀x A(x) → ∃y B(y)) is a theorem, so is ∃y (∀x A(x) → B(y)).

Both rules are also admissible for HA. The corresponding implications (MP and IP respectively), which are not provable intuitionistically, are verified by Gödel's "Dialectica" interpretation of HA. So is the implication (CT) corresponding to one of the most interesting admissible rules of Heyting arithmetic, let us call it the Church-Kleene Rule:

If ∀x ∃y A(x, y) is a closed theorem of HA then there is a number n such that, provably in HA, the partial recursive function with Gödel number n is total and maps each x to a y satisfying A(x, y) (and moreover A(x,y) is provable, where x is the numeral for the natural number x and y is the numeral for y).

Combining Markov's Rule with the negative translation gives the result that classical and intuitionistic arithmetic prove the same formulas of the form ∀x ∃y A(x, y) where A(x, y) is quantifier-free. In general, if A(x, y) is provably decidable in HA and if ∀x ∃y A(x, y) is a closed theorem of classical arithmetic PA, the conclusion of the Church-Kleene Rule holds even in intuitionistic arithmetic. For if HA proves ∀x ∀y (A(x,y) ∨ ¬A(x,y)) then by the Church-Kleene Rule the characteristic function of A(x,y) has a Gödel number m, provably in HA; so HA proves ∀x ∃y A(x,y) ↔ ∀x ∃y ∃z B(m,x,y,z) where B is quantifier-free, and the adjacent existential quantifiers can be contracted in HA. It follows that HA and PA have the same provably recursive functions.

Here is a proof that the rule "If ∀x (A ∨ B(x)) is a theorem, so is A ∨ ∀x B(x)" (where x is not free in A) is not admissible for HA, if HA is consistent. Gödel numbering provides a quantifier-free formula G(x) which (numeralwise) expresses the predicate "x is the code of a proof in HA of (0 = 1)." By intuitionistic logic with the decidability of quantifier-free arithmetical formulas, HA proves ∀x(∃yG(y) ∨ ¬G(x)). However, if HA proved ∃yG(y) ∨ ∀x¬G(x) then by the disjunction property, HA must prove either ∃yG(y) or ∀x¬G(x). The first case is impossible, by the existence property with the consistency assumption and the fact that HA proves all true quantifier-free sentences.

But the second case is also impossible, by Gödel's second incompleteness theorem, since $\forall x \neg G(x)$ expresses the consistency of HA.

Basic Semantics

Kripke Semantics for Intuitionistic Logic

Intuitionistic systems have inspired a variety of interpretations, including Beth's tableaus, Rasiowa and Sikorski's topological models, formulas-as-types, Kleene's recursive realizabilities, the Kleene and Aczel slashes, and models based on sheafs and topoi. Kripke's possible-world semantics, with respect to which intuitionistic predicate logic is complete and consistent, most resembles classical model theory.

A Kripke structure K for L consists of a partially ordered set K of nodes and a domain function D assigning to each node k in K an inhabited set D(k), such that if $k \leq k'$, then $D(k) \subseteq D(k')$. In addition K has a forcing relation determined as follows.

For each node k let L(k) be the language extending L by new constants for all the elements of D(k). To each node k and each 0-ary predicate letter (each proposition letter) P, either assign f(P, k) = true or leave f(P, k) undefined, consistent with the requirement that if $k \leq k'$ and f(P, k) = true then f(P, k') = true also. Say that

    k forces P if and only if f(P, k) = true.

To each node k and each (n+1)-ary predicate letter Q(...), assign a (possibly empty) set T(Q, k) of (n+1)-tuples of elements of D(k) in such a way that if $k \leq k'$ then $T(Q, k) \subseteq T(Q, k')$. Say that

    k forces $Q(d_0,...,d_n)$ if and only if $(d_0,...,d_n) \in T(Q, k)$.

Now define forcing for compound sentences of L(k) inductively as follows:

- k forces (A & B) if k forces A and k forces B.

- k forces (A ∨ B) if k forces A or k forces B.

- k forces (A → B) if, for every $k' \geq k$, if k' forces A then k' forces B.

- k forces ¬A if for no $k' \geq k$ does k' force A.

- k forces ∀xA(x) if for every $k' \geq k$ and every $d \in D(k')$, k' forces A(d).

- k forces ∃xA(x) if for some $d \in D(k)$, k forces A(d).

Any such forcing relation is consistent and monotone:

- for no sentence A and no k does k force both A and ¬A.

- if $k \leq k'$ and k forces A then k' forces A.

Kripke's Soundness and Completeness Theorems establish that a sentence of L is provable in intuitionistic predicate logic if and only if it is forced by every node of every Kripke structure. Thus to show that $(\neg\forall x \neg P(x) \rightarrow \exists x P(x))$ is intuitionistically unprovable, it is enough to consider a Kripke structure with K = {k, k'}, k < k', D(k) = D(k') = {0}, T(P, k) empty but T(P, k') = {0}.

And to show the converse is intuitionistically provable (without actually exhibiting a proof), one only needs the consistency and monotonicity properties of arbitrary Kripke models, with the definition of forcing.

Kripke models for languages with equality may interpret = at each node by an arbitrary equivalence relation, subject to monotonicity. For applications to intuitionistic arithmetic, normal models (those in which equality is interpreted by identity at each node) suffice because equality of natural numbers is decidable.

Propositional Kripke semantics is particulary simple, since an arbitrary propositional formula is intuitionistically provable if and only if it is forced by the root of every Kripke model whose frame (the set K of nodes together with their partial ordering) is a finite tree with a least element (the root). For example, the Kripke model with $K = \{k, k', k''\}$, $k < k'$ and $k < k''$, and with P true only at k', shows that both $P \vee \neg P$ and $\neg P \vee \neg\neg P$ are unprovable in IPC.

Each terminal node or leaf of a Kripke model is a classical model, because a leaf forces every formula or its negation. Only those proposition letters which occur in a formula E, and only those nodes k' such that $k \leq k'$, are relevant to deciding whether or not k forces E. Such considerations allow us to associate effectively with each formula E of L(IPC) a finite class of finite Kripke structures which will include a countermodel to E if one exists. Since the class of all theorems of IPC is recursively enumerable, we conclude that

> IPC is effectively decidable. There is a recursive procedure which determines, for each propositional formula E, whether or not E is a theorem of IPC, concluding with either a proof of E or a Kripke countermodel.

The decidability of IPC was first obtained by Gentzen in 1933 as an immediate corollary of his Hauptsatz. The undecidability of IQC follows from the undecidability of CQC by the negative interpretation.

Familiar non-intuitionistic logical schemata correspond to structural properties of Kripke models, for example

- DNS holds in every Kripke model with finite frame.

- $(A \rightarrow B) \vee (B \rightarrow A)$ holds in every Kripke model with linearly ordered frame. Conversely, every propositional formula which is not derivable in IPC + $(A \rightarrow B) \vee (B \rightarrow A)$ has a Kripke countermodel with linearly ordered frame.

- If x is not free in A then $(\forall x(A \vee B(x)) \rightarrow (A \vee \forall x\, B(x)))$ holds in every Kripke model K with constant domain (so that $D(k) = D(k')$ for all k, k' in K). The same is true for MP.

Kripke models are a powerful tool for establishing properties of intuitionistic formal systems; Troelstra and van Dalen, Smorynski, de Jongh and Smorynski, Ghilardi and Iemhoff. Following Gödel, Kreisel argued that Kripke-completeness of intuitionistic logic entailed Markov's Principle. By modifying the definition of a Kripke model to allow "exploding nodes" which force every sentence, Veldman found an intuitionistic completeness proof avoiding (the informal use of) MP.

Realizability Semantics for Heyting Arithmetic

One way to implement the B-H-K explanation of intuitionistic truth for arithmetic is to associate with each sentence E of HA some collection of numerical codes for algorithms which could establish the constructive truth of E. Following Kleene , a number e realizes a sentence E of the language of arithmetic by induction on the logical form of E:

- e realizes $(r = t)$, if $(r = t)$ is true.

- e realizes (A & B), if e codes a pair (f,g) such that f realizes A and g realizes B.

- e realizes A∨B, if e codes a pair (f,g) such that if $f = 0$ then g realizes A, and if $f > 0$ then g realizes B.

- e realizes A→B, if, whenever f realizes A, then the eth partial recursive function is defined at f and its value realizes B.

- e realizes ¬A, if no f realizes A.

- e realizes ∀x A(x), if, for every n, the eth partial recursive function is defined at n and its value realizes A(n).

- e realizes ∃x A(x), if e codes a pair (n,g) and g realizes A(n).

An arbitrary formula is realizable if some number realizes its universal closure. Observe that not both A and ¬A are realizable, for any formula A. The fundamental result is;

Nelson's Theorem: If A is derivable in HA from realizable formulas F, then A is realizable.

Some nonintuitionistic principles can be shown to be realizable. For example, Markov's Principle (for decidable formulas) can be expressed by the schema

(MP) ∀x(A(x) ∨ ¬A(x)) & ¬∀x¬A(x) → ∃x A(x).

Although unprovable in HA, MP is realizable by an argument which uses Markov's Principle informally. But realizability is a fundamentally nonclassical interpretation. In HA it is possible to express an axiom of recursive choice CT (for "Church's Thesis"), which contradicts LEM but is (constructively) realizable. Hence by Nelson's Theorem, HA + MP + CT is consistent.

Kleene used a variant of number-realizability to prove HA satisfies the Church-Kleene Rule; the same argument works for HA with MP and/or CT. In Kleene and Vesley and Kleene , functions replace numbers as realizing objects, establishing the consistency of formalized intuitionistic analysis and its closure under a second-order version of the Church-Kleene Rule.

De Jongh combined realizability with Kripke modeling to show that intuitionistic predicate logic is arithmetically complete for HA. If, to each n-place predicate letter P(...), a formula f(P) of L(HA) with n free variables is assigned, and if the formula f(A) of L(HA) comes from the formula A of L by replacing each prime formula $P(x_1,..., x_n)$ by $f(P)(x_1,..., x_n)$, then f(A) is called an arithmetical substitution instance of A. A uniform assignment of simple existential formulas to predicate letters suffices to prove;

De Jongh's Theorem: If a sentence A of the language L is not provable in IQC, then some arithmetical substitution instance of A is not provable in HA.

For example, if P(x, y) expresses "x codes a proof in HA of the formula with code y," then $\forall y$ ($\exists x$ P(x, y) $\vee$ $\neg \exists x$ P(x, y)) is unrealizable, hence unprovable in HA, and so is its double negation. (The proof of de Jongh's Theorem for IPC does not need realizability. As an example, Rosser's form of Gödel's Incompleteness Theorem provides a sentence C of L(HA) such that PA proves neither C nor $\neg$C, so by the disjunction property HA cannot prove (C $\vee$ $\neg$C).)

Without claiming that number-realizability coincides with intuitionistic arithmetical truth, Nelson observed that for each formula A of L(HA) the predicate "y realizes A" can be expressed in HA by another formula (abbreviated "y re A"), and the schema A $\leftrightarrow$ $\exists y$ (y re A) is consistent with HA. Troelstra showed that HA + (A $\leftrightarrow$ $\exists y$ (y re A)) is equivalent to HA + ECT, where ECT is a strengthened form of CT. In HA + MP + ECT, which Troelstra considers to be a formalization of Russian recursive mathematics, every formula of the form (y re A) has an equivalent "classical" prenex form A'(y) consisting of a quantifier-free subformula preceded by alternating "classical" quantifiers of the forms $\neg\neg\exists x$ and $\forall z\neg\neg$, and so $\exists y$ A'(y) is a kind of prenex form of A.

Subintuitionistic and Superintuitionistic Logics

At present there are several other entries treating intuitionistic logic in various contexts, but a general treatment of intermediate logics appears to be lacking so a brief one is included here. A subintuitionistic propositional logic can be obtained from IPC by restricting the language, or weakening the logic, or both. An extreme example of the first is RN, intuitionistic logic with a single propositional variable P, which is named after its discoverers Rieger and Nishimura . RN is characterized by the Rieger-Nishimura lattice of infinitely many nonequivalent formulas F_n such that every formula whose only propositional variable is P is equivalent by intuitionistic logic to some F_n. Nishimura's version is

- $F_\infty = P \rightarrow P.$

- $F_0 = P \,\&\, \neg\, P.$

- $F_1 = P.$

- $F_2 = \neg\, P.$

- $F_{2n+3} = F_{2n+1} \vee F_{2n+2}.$

- $F_{2n+4} = F_{2n+3} \rightarrow F_{2n+1}.$

In RN neither F_{2n+1} nor F_{2n+2} implies the other; but F_{2n} implies F_{2n+1}, and F_{2n+1} implies each of F_{2n+3} and F_{2n+4}.

Fragments of IPC missing one or more logical connective restrict the language and incidentally the logic, since the intuitionistic connectives &, $\vee$, $\rightarrow$, $\neg$ are logically independent over IPC. Rose proved that the implicationless fragment (without $\rightarrow$) is complete with respect to realizability, in the sense that if every arithmetical substitution instance of a propositional formula E without $\rightarrow$ is (number)-realizable then E is a theorem of IPC. This result contrasts with;

Rose's Theorem: IPC is incomplete with respect to realizability. Let F be the propositional formula

$$((\neg\,\neg\, D \rightarrow D) \rightarrow (\neg\,\neg\, D \vee \neg\, D)) \rightarrow (\neg\,\neg\, D \vee \neg\, D)$$

where D is $(\neg P \vee \neg Q)$ and P, Q are prime. Every arithmetical substitution instance of F is realizable (using classical logic), but F is not provable in IPC.

It follows that IPC is arithmetically incomplete for HA + ECT.

An intermediate propositional logic is any consistent collection of propositional formulas containing all the axioms of IPC and closed under modus ponens and substitution of arbitrary formulas for proposition letters. Each intermediate propositional logic is contained in CPC. Some particular intermediate propositional logics, characterized by adding one or more classically correct but intuitionistically unprovable axiom schemas to IPC, have been studied extensively.

One of the simplest intermediate propositional logics is the Gödel-Dummett logic LC, obtained by adding to IPC the schema $(A \rightarrow B) \vee (B \rightarrow A)$ which is valid on all and only those Kripke frames in which the partial order of the nodes is linear. Gödel used an infinite sequence of successively stronger intermediate logics to show that IPC has no finite truth-table interpretation. For each positive integer n, let G_n be LC plus the schema $(A_1 \rightarrow A_2) \vee ... \vee (A_1 \& ... \& A_n \rightarrow A_{n+1})$. Then G_n is valid on all and only those linearly ordered Kripke frames with no more than n nodes.

The Kreisel-Putnam logic KP, obtained by adding to IPC the schema $(\neg A \rightarrow B \vee C) \rightarrow ((\neg A \rightarrow B) \vee (\neg A \rightarrow C))$, has the disjunction property but does not satisfy all the Visser rules. The intermediate logic obtained by adding the schema $((\neg \neg D \rightarrow D) \rightarrow (D \vee \neg D)) \rightarrow (\neg \neg D \vee \neg D)$, corresponding to Rose's counterexample, to IPC also has the disjunction property. Iemhoff proved that IPC is the only intermediate propositional logic with the disjunction property which is closed under all the Visser rules. Iemhoff and Metcalfe developed a formal calculus for generalized admissibility for IPC and some intermediate logics.

An intermediate propositional logic L is said to have the finite frame property if there is a class of finite frames on which the Kripke-valid formulas are exactly the theorems of L. Many intermediate logics, including LC (the class of finite linear frames) and KP, have this property. De Jongh, Verbrugge and Visser proved that every intermediate logic L with the finite frame property is the propositional logic of HA(L), that is, the class of all formulas in the language of IPC all of whose arithmetical substitution instances are provable in the logical extension of HA by L.

Some intermediate predicate logics, extending IQC and closed under substitution, are IQC + DNS (Section 4.1), IQC + MP, IQC + MP + IP, and the intuitionistic logic of constant domains CD obtained by adding to IQC the schema $\forall x(A \vee B(x)) \rightarrow (A \vee \forall x\, B(x))$ for all formulas A, B(x) with x not occurring free in A. Mints, Olkhovikov and Urquhart [2012, Other Internet Resources] showed that CD does not have the interpolation property, refuting earlier published proofs by other authors.

Axiomatic Systems

In mathematics and set theory, an axiomatic system is any set of specified axioms from which some or all of those axioms can be used, in conjunction along with derivation rules or procedures, to logically derive theorems. A mathematical theory or set theory consists of an axiomatic system and all its derived theorems. An axiomatic system that is completely described is a special kind of formal system; usually, however, the effort towards complete formalization brings diminishing returns in certainty and a lack of readability for humans. Therefore discussion of axiomatic systems is normally only semi-formal. A formal theory typically means an axiomatic system, for example formulated within model theory. A formal proof is a complete rendition of a mathematical or set-theoritical proof within a formal system.

Properties

An axiomatic system is said to be *consistent* if it lacks *contradiction* (i.e. it is not possible to derive both a statement and its negation from the system's axioms).

In an axiomatic system, an axiom is called *independent* if it is not a theorem that can be derived from other axioms in the system. A system will be called *independent* if each of its underlying axioms is independent.

The most important criterion for assessment of an axiomatic system is that particular system's consistency. Inconsistency in an axiomatic system is universally regarded as being a fatal flaw for that system.

Independence is also a desireable property, but its lack is not a fatal flaw. Lack of independence means that the system has redundancy in its axioms, meaning that one or more of its axioms is not needed. This is usually considered to be a flaw because reducing the number of axioms of a system to the minimum necessary for deriving all the needed or desired theorems of that system is considered to be a virtue, because axioms are unproved and unprovable; having as little of that as possible means that as few unproved assumptions as possible are being made in that system.

An axiomatic system will be called *complete* if for every statement, either itself or its negation, is derivable in that system. This is very difficult to achieve, however, and as shown by the combined works of Gödel and Coen, impossible for axiomatic systems involving infinite sets. So, along with consistency, relative consistency is also the mark of a worthwhile axiom system. This occurs when the undefined terms of a first axiom system are given definitions from a second, such that the axioms of the first are theorems of the second.

A good example is the relative consistency of neutral geometry, or *absolute geometry*, with respect to the theory of the real number system. Lines and points are undefined terms in absolute geometry, but assigned meanings in the theory of real numbers in a way that is consistent with both axiom systems.

Models

A *model* for an axiomatic system is a well-defined set, which assigns meaning for the undefined terms presented in the system, in a manner that is correct with the relations defined in the system. The existence of a *concrete model* proves the *consistency* of a system.

A model is called *concrete* if the meanings assigned are objects and relations from the real world, as opposed to an *abstract model* which is based on other axiomatic systems. The first axiomatic system was Euclidean geometry.

Models can also be used to show the *independence* of an axiom in the system. By constructing a valid model for a subsystem without a specific axiom, we show that the omitted axiom is *independent* if its correctness does not necessarily follow from the subsystem.

Two models are said to be isomorphic if a one-to-one correspondence can be found between their elements, in a manner that preserves their relationship. An axiomatic system for which every model is isomorphic to another is called *categorial* (sometimes *categorical*), and the property of *categoriality (categoricity)* ensures the *completeness* of a system.

Axiomatic Method

The axiomatic method is often discussed as if it were a unitary approach, or uniform procedure. With the example of Euclid to appeal to, it was indeed treated that way for many centuries. Up until the beginning of the nineteenth century it was generally assumed in European mathematics and philosophy (for example in Spinoza's work) that the heritage of Greek mathematics represented the highest standard of intellectual finish (development *more geometrico*, in the style of the geometers).

That approach, in which axioms were supposed to be *self-evident* and thus indisputable, was swept away during the course of the nineteenth century. One important episode in this was the development of Non-Euclidean geometry, based on denial of Euclid's parallel postulate (or axiom). It was found that consistent geometries can be constructed by denying that postulate, taking as an axiom that more than one parallel to a given line can be drawn through a point outside that line, or a different axiom that no parallel can be drawn—both of those result in different and consistent geometric systems that may or may not be applicable to an experienced world.

Other challenges to the supposed self-evidence of axioms came from the foundations of real analysis, from Georg Cantor's set theory, and from the failure of Frege's work on foundations. Russell was able to derive a paradox—a kind of contradiction—from Frege's axioms for set theory, thus showing that Frege's axiomatic system was not consistent, and this showed that the supposed self-evidence of Frege's axioms was mistaken.

Another challenge came from David Hilbert's 'new' use of axiomatic method as a research tool. For example, group theory was first put on an axiomatic basis towards the end of that century. Once the axioms were clarified (that inverse elements should be required, for example), the subject could proceed autonomously without reference to the transformation group's origins of those studies.

Therefore, there are at least three 'modes' of axiomatic method current in mathematics, and in the fields it influences. In caricature, possible attitudes are:

1. Accept my axioms and you must accept their consequences;

2. I reject one of your axioms and accept extra models;

3. My set of axioms defines a research program.

The first case is the classic deductive method. The second goes by the slogan *be wise, generalise*; it may go along with the assumption that concepts can or should be expressed at some intrinsic 'natural level of generality'. The third was very prominent in the mathematics of the twentieth century, in particular in subjects based around homological algebra.

It is easy to see that the axiomatic method has limitations outside mathematics and set theory. For example, in political philosophy axioms that lead to unacceptable conclusions are likely to be rejected wholesale; so that no one really assents to version 1.

Axiomatic Systems in Science

Historically, the most important purpose of an axiom system was to reach an overview of some science or part of science. Euclid succeeded in doing that for geometry. Beyond its successful use in set theory and mathematics, there have been attempts to use the axiomatic method in physics (by Ludwig Boltzmann, Heinrich Hertz, and some members of the Vienna Circle), biology (by J. H. Woodger), quantum mechanics (by Günther Ludwig), and possibly other sciences.

Those attempts have been at best only partly successful. If such efforts at axiomatization for sciences were to be successful, this would make it possible to study these sciences simply by drawing logical inferences from the axioms, without needing any new empirical input. When conclusions are drawn from general scientific laws or principles that method is in fact employed, empirical testing of such theoretical derivations is still always needed. Thus, the axiomatization and formalization of the system is incomplete and does not solve the problem of whether the system yields actual scientific knowledge.

Issues

Not every consistent body of propositions can be captured by a describable collection of axioms. Call a collection of axioms recursive if a computer program can recognize whether a given proposition in the language is an axiom. Gödel's First Incompleteness Theorem then tells us that there are certain consistent bodies of propositions with no recursive axiomatization. Typically, the computer can recognize the axioms and logical rules for deriving theorems, and the computer can recognize whether a proof is valid, but to determine whether a proof exists for a statement is only soluble by "waiting" for the proof or disproof to be generated. The result is that one will not know which propositions are theorems and the axiomatic method breaks down. An example of such a body of propositions is the theory of the natural numbers. The Peano Axioms (described below) thus only partially axiomatize this theory.

In practice, not every proof is traced back to the axioms. At times, it is not clear which collection of axioms a proof appeals to. For example, a number-theoretic statement might be expressible in the language of arithmetic (i.e. the language of the Peano Axioms) and a proof might be given that appeals to topology or complex analysis. It might not be immediately clear whether another proof can be found that derives itself solely from the Peano Axioms.

Any more-or-less arbitrarily chosen system of axioms is the basis of some mathematical theory, but such an arbitrary axiomatic system will not necessarily be free of contradictions, and even if it is, it is not likely to shed light on anything. Philosophers of mathematics sometimes assert that mathematicians choose axioms "arbitrarily", but it is possible that although they may appear arbitrary when viewed only from the point of view of the canons of deductive logic, that appearance is due to a limitation on the purposes that deductive logic serves.

Example: The Peano axiomatization of natural numbers

The mathematical system of natural numbers 0, 1, 2, 3, 4, … is based on an axiomatic system first written down by the mathematician Peano in 1889. He chose the axioms, in the language of a single unary function symbol S (short for "successor"), for the set of natural numbers to be:

- There is a natural number 0.

- Every natural number a has a successor, denoted by Sa.

- There is no natural number whose successor is 0.

- Distinct natural numbers have distinct successors: if $a \neq b$, then $Sa \neq Sb$.

- If a property is possessed by 0 and also by the successor of every natural number it is possessed by, then it is possessed by all natural numbers ("*Induction axiom*").

Axiomatization

In mathematics, axiomatization is the formulation of a system of statements (i.e. axioms) that relate a number of primitive terms in order that a consistent body of propositions may be derived deductively from these statements. Thereafter, the proof of any proposition should be, in principle, traceable back to these axioms.

Hilbert System

A *Hilbert system* is a style (formulation) of deductive system that emphasizes the role played by the axioms in the system. Typically, a Hilbert system has many axiom schemes, but only a few, sometimes one, rules of inference. As such, a Hilbert system is also called an axiom system. Below we list three examples of axiom systems in mathematical logic:

- (intuitionistic propositional logic)

    -axiom schemes:

    i. $A \to (B \to A)$

    ii. $(A \to (B \to C)) \to ((A \to B) \to (A \to C))$

    iii. $A \to A \lor B$

    iv. $B \to A \lor B$

    v. $(A \to C) \to ((B \to C) \to (A \lor B \to C))$

    vi. $A \land B \to A$

    vii. $A \land B \to B$

viii. $A\to(B\to(A\wedge B))$

ix. $\bot\to A$

-rule of inference: (modus ponens): from $A\to B$ and A , we may infer B

- (classical predicate logic without equality)

    -axiom schemes:

    i.   all of the axiom schemes above, and

    ii.   law of double negation: $\neg(\neg A)\to A$

    iii. $\forall xA\to A[x/y]$

    iv. $\forall x(A\to B)\to(A\to\forall yB[x/y])$

In the last two axiom schemes, we require that y is free for x in A , and in the last axiom scheme, we also require that x does not occur free in A .

    -rules of inference:

    i.   modus ponens, and

    ii.   generalization: from A , we may infer $\forall yA[x/y]$, where y is free for x in A

- (S4 modal propositional logic)

    – axiom schemes:

    i.   all of the axiom schemes in intuitionistic propositional logic, as well as the law of double negation, and

    ii.   Axiom K, or the normality axiom: $\Box(A\to B)\to(\Box A\to\Box B)$

    iii.   Axiom T: $\Box A\to A$

    iv.   Axiom 4: $\Box A\to\Box(\Box A)$

    – rules of inference:

    i.   modus ponens, and

    ii.   necessitation: from A , we may infer $\Box A$

where A,B,C above are well-formed formulas, x,y are individual variables, and $\to,\vee,\wedge$ are binary, $\Box$ unary, and $\bot$ nullary logical connectives in the respective logical systems. The connective $\neg$ may be defined as $\neg A:=A\to\bot$ for any formula A .

References

- Barwise, Jon (1977); "An Introduction to First-Order Logic", in Barwise, Jon, ed. (1982). Handbook of Mathematical Logic. Studies in Logic and the Foundations of Mathematics. Amsterdam, NL: North-Holland. ISBN 978-0-444-86388-1

- Axiomatic-systems: newworldencyclopedia.org, Retrieved 30 June 2018

- Avigad, Jeremy; Donnelly, Kevin; Gray, David; and Raff, Paul (2007); "A formally verified proof of the prime number theorem", ACM Transactions on Computational Logic, vol. 9 no. 1 doi:10.1145/1297658.1297660

- Non-standard-arithmetic, model-theory: settheory.net, Retrieved 09 July 2018

- Gamut, L. T. F. (1991); Logic, Language, and Meaning, Volume 2: Intensional Logic and Logical Grammar, Chicago, Illinois: University of Chicago Press, ISBN 0-226-28088-8

- First-order-logic: whatis.techtarget.com, Retrieved 29 June 2018

- Rautenberg, Wolfgang (2010), A Concise Introduction to Mathematical Logic (3rd ed.), New York, NY: Springer Science+Business Media, doi:10.1007/978-1-4419-1221-3, ISBN 978-1-4419-1220-6

4

Formal Logic Systems

Mathematical logic can be of different forms, like infinitary logic, propositional logic, categorical logic, classical and non-classical logic, algebraic logic, etc. This chapter has been carefully written to provide an easy understanding of the different types of mathematical logic along with their principles and applications.

Propositional Logic

Propositional logic, also known as sentential logic and statement logic, is the branch of logic that studies ways of joining and/or modifying entire propositions, statements or sentences to form more complicated propositions, statements or sentences, as well as the logical relationships and properties that are derived from these methods of combining or altering statements. In propositional logic, the simplest statements are considered as indivisible units, and hence, propositional logic does not study those logical properties and relations that depend upon parts of statements that are not themselves statements on their own, such as the subject and predicate of a statement. The most thoroughly researched branch of propositional logic is classical truth-functional propositional logic, which studies logical operators and connectives that are used to produce complex statements whose truth-value depends entirely on the truth-values of the simpler statements making them up, and in which it is assumed that every statement is either true or false and not both. However, there are other forms of propositional logic in which other truth-values are considered, or in which there is consideration of connectives that are used to produce statements whose truth-values depend not simply on the truth-values of the parts, but additional things such as their necessity, possibility or relatedness to one another.

The Truth Value of a proposition is True(denoted as T) if it is a true statement, and False(denoted as F) if it is a false statement.

For example:

1. The sun rises in the East and sets in the West.

2. $1 + 1 = 2$.

3. 'b' is a vowel.

All of the above sentences are propositions, where the first two are Valid(True) and the third one is Invalid(False).

Some sentences that do not have a truth value or may have more than one truth value are not propositions.

For example:

1. What time is it?

2. Go out and play.

3. x + 1 = 2.

The above sentences are not propositions as the first two do not have a truth value, and the third one may be true or false.

To represent propositions, propositional variables are used. By Convention, these variables are represented by small alphabets such as p,q,r,s .

The area of logic which deals with propositions is called propositional calculus or propositional logic.

It also includes producing new propositions using existing ones. Propositions constructed using one or more propositions are called compound propositions. The propositions are combined together using Logical Connectives or Logical Operators.

Truth Table

Since we need to know the truth value of a proposition in all possible scenarios, we consider all the possible combinations of the propositions which are joined together by Logical Connectives to form the given compound proposition. This compilation of all possible scenarios in a tabular format is called a truth table.

Most Common Logical Connectives-

1. Negation – If p is a proposition, then the negation of p is denoted by $\neg p$, which when translated to simple English means-

"It is not the case that p" or simply "not p".

The truth value of $\neg p$ is the opposite of the truth value of p.

The truth table of $\neg p$ is-

p	$\neg p$
T	F
F	T

Example,

The negation of "It is raining today", is "It is not the case that is raining today" or simply "It is not raining today".

2. Conjunction – For any two propositions p and q, their conjunction is denoted by $p \wedge q$, which means "p and q". The conjuction $p \wedge q$ is True when both p and q are True, otherwise False. The truth table of $p \wedge q$ is-

p	q	$p \wedge q$
T	T	T
T	F	F
F	T	F
F	F	F

Example,

The conjunction of the propositions p – "Today is Friday" and q – "It is raining today", $p \wedge q$ is "Today is Friday and it is raining today". This proposition is true only on rainy Fridays and is false on any other rainy day or on Fridays when it does not rain.

3. Disjunction – For any two propositions p and q, their disjunction is denoted by $p \vee q$, which means "p or q". The disjuction $p \vee q$ is True when either p or q is True, otherwise False. The truth table of $p \vee q$ is-

p	q	$p \vee q$
T	T	T
T	F	T
F	T	T
F	F	F

Example,

The disjunction of the propositions p – "Today is Friday" and q – "It is raining today", $p \vee q$ is "Today is Friday or it is raining today". This proposition is true on any day that is a Friday or a rainy day(including rainy Fridays) and is false on any day other than Friday when it also does not rain.

4. Exclusive Or – For any two propositions p and q, their exclusive or is denoted by $p \oplus q$, which means "either p or q but not both". The exclusive or $p \oplus q$ is True when either   or q is True, and False when both are true or both are false.

The truth table of $p \oplus q$ is-

p	q	$p \oplus q$
T	T	F

T	F	T
F	T	T
F	F	F

Example,

The exclusive or of the propositions p – "Today is Friday" and q – "It is raining today", $p \oplus q$ is "Either today is Friday or it is raining today, but not both". This proposition is true on any day that is a Friday or a rainy day(not including rainy Fridays) and is false on any day other than Friday when it does not rain or rainy Fridays.

5. Implication – For any two propositions p and q, the statement "if p then q" is called an implication and it is denoted by $p \rightarrow q$.

In the implication $p \rightarrow q$, p is called the hypothesis or antecedent or premise and q is called the conclusion or consequence.

The implication is $p \rightarrow q$ is also called a conditional statement.

The implication is false when p is true and q is false otherwise it is true. The truth table of $p \rightarrow q$ is-

p	q	$p \rightarrow q$
T	T	
T	F	F
F	T	T
F	F	T

You might wonder that why is $p \rightarrow q$ true when p is false. This is because the implication guarantees that when p and q are true then the implication is true. But the implication does not guarantee anything when the premise p is false. There is no way of knowing whether or not the implication is false since p did not happen.

This situation is similar to the "Innocent until proven Guilty" stance, which means that the implication $p \rightarrow q$ is considered true until proven false. Since we cannot call the implication $p \rightarrow q$ false when p is false, our only alternative is to call it true.

This follows from the Explosion Principle which says-

"A False statement implies anything"

Conditional statements play a very important role in mathematical reasoning, thus a variety of terminology is used to express $p \rightarrow q$, some of which are listed below.

"if p, then q"
"p is sufficient for q"
"q when p"
"a necessary condition for p is q"
"p only if q"
"q unless $\neg p$"
"q follows from p"

Example,

"If it is Friday then it is raining today" is a proposition which is of the form $p \rightarrow q$. The above proposition is true if it is not Friday(premise is false) or if it is Friday and it is raining, and it is false when it is Friday but it is not raining.

6. Biconditional or Double Implication – For any two propositions p and q, the statement "p if and only if(iff) q" is called a biconditional and it is denoted by $p \leftrightarrow q$.

The statement $p \leftrightarrow q$ is also called a bi-implication.

$p \leftrightarrow q$ has the same truth value as $(p \rightarrow q) \wedge (q \rightarrow p)$

The implication is true when p and q have same truth values, and is false otherwise. The truth table of $p \leftrightarrow q$ is-

p	q	$p \leftrightarrow q$
T	T	T
T	F	F
F	T	F
F	F	T

Some other common ways of expressing $p \leftrightarrow q$ are-

"p is necessary and sufficient for q"
"if p then q, and conversely"
"p iff q"

Example,

"It is raining today if and only if it is Friday today." is a proposition which is of the form $p \leftrightarrow q$. The above proposition is true if it is not Friday and it is not raining or if it is Friday and it is raining, and it is false when it is not Friday or it is not raining.

Tautologies

A Tautology is a formula which is always true for every value of its propositional variables.

Example – Prove $[(A{\rightarrow}B){\wedge}A]{\rightarrow}B$ is a tautology

The truth table is as follows –

A	B	A → B	(A → B) ∧ A	[(A → B) ∧ A] → B
True	True	True	True	True
True	False	False	False	True
False	True	True	False	True
False	False	True	False	True

As we can see every value of $[(A{\rightarrow}B) \wedge A] {\rightarrow}B$ is "True", it is a tautology.

Contradictions

A Contradiction is a formula which is always false for every value of its propositional variables.

Example – Prove $(A{\vee}B) \wedge [({\neg}A) \wedge ({\neg}B)]$ is a contradiction

The truth table is as follows –

A	B	A ∨ B	¬ A	¬ B	(¬ A) ∧ (¬ B)	(A ∨ B) ∧ [(¬ A) ∧ (¬ B)]
True	True	True	False	False	False	False
True	False	True	False	True	False	False
False	True	True	True	False	False	False
False	False	False	True	True	True	False

As we can see every value of $(A{\vee}B) \wedge [({\neg}A) \wedge ({\neg}B)]$ is "False", it is a contradiction.

Contingency

A Contingency is a formula which has both some true and some false values for every value of its propositional variables.

Example – Prove $(A{\vee}B) \wedge ({\neg}A)$ a contingency

The truth table is as follows –

A	B	A ∨ B	¬ A	(A ∨ B) ∧ (¬ A)
True	True	True	False	False
True	False	True	False	False
False	True	True	True	True
False	False	False	True	False

As we can see every value of $(A{\vee}B) \wedge ({\neg}A)$ has both "True" and "False", it is a contingency.

Propositional Equivalences

Two statements X and Y are logically equivalent if any of the following two conditions hold –

- The truth tables of each statement have the same truth values.

- The bi-conditional statement $X \Leftrightarrow Y$ is a tautology.

Example – Prove $\neg(A \lor B)$ and $[(\neg A) \land (\neg B)]$ are equivalent

Testing by 1st method (Matching truth table)

A	B	A ∨ B	¬ (A ∨ B)	¬ A	¬ B	[(¬ A) ∧ (¬ B)]
True	True	True	False	False	False	False
True	False	True	False	False	True	False
False	True	True	False	True	False	False
False	False	False	True	True	True	True

Here, we can see the truth values of $\neg(A \lor B)$ and $[(\neg A) \land (\neg B)]$ are same, hence the statements are equivalent.

Testing by 2nd method (Bi-conditionality)

A	B	¬ (A ∨ B)	[(¬ A) ∧ (¬ B)]	[¬ (A ∨ B)] ⇔ [(¬ A) ∧ (¬ B)]
True	True	False	False	True
True	False	False	False	True
False	True	False	False	True
False	False	True	True	True

As $[\neg(A \lor B)] \Leftrightarrow [(\neg A) \land (\neg B)]$ is a tautology, the statements are equivalent.

Inverse, Converse and Contra-positive

Implication / if-then ($\rightarrow$) is also called a conditional statement. It has two parts –

- Hypothesis, p
- Conclusion, q

As mentioned earlier, it is denoted as $p \rightarrow q$.

Example of Conditional Statement – "If you do your homework, you will not be punished." Here, "you do your homework" is the hypothesis, p, and "you will not be punished" is the conclusion, q.

Inverse – An inverse of the conditional statement is the negation of both the hypothesis and the conclusion. If the statement is "If p, then q", the inverse will be "If not p, then not q". Thus the inverse of $p \rightarrow q$ is $\neg p \rightarrow \neg q$.

Example – The inverse of "If you do your homework, you will not be punished" is "If you do not do your homework, you will be punished."

Converse – The converse of the conditional statement is computed by interchanging the hypothesis and the conclusion. If the statement is "If p, then q", the converse will be "If q, then p". The converse of $p{\rightarrow}q$ is $q{\rightarrow}p$.

Example – The converse of "If you do your homework, you will not be punished" is "If you will not be punished, you do your homework".

Contra-positive – The contra-positive of the conditional is computed by interchanging the hypothesis and the conclusion of the inverse statement. If the statement is "If p, then q", the contra-positive will be "If not q, then not p". The contra-positive of $p{\rightarrow}q$ is $\neg q{\rightarrow}\neg p$.

Example – The Contra-positive of " If you do your homework, you will not be punished" is "If you are punished, you do your homework".

Duality Principle

Duality principle states that for any true statement, the dual statement obtained by interchanging unions into intersections (and vice versa) and interchanging Universal set into Null set (and vice versa) is also true. If dual of any statement is the statement itself, it is said self-dual statement.

Example – The dual of $(A{\cap}B){\cup}C$ is $(A{\cup}B){\cap}C$

Normal Forms

We can convert any proposition in two normal forms –

- Conjunctive normal form
- Disjunctive normal form

Conjunctive Normal Form

A compound statement is in conjunctive normal form if it is obtained by operating AND among variables (negation of variables included) connected with ORs. In terms of set operations, it is a compound statement obtained by Intersection among variables connected with Unions.

Examples

- $(A{\vee}B){\wedge}(A{\vee}C){\wedge}(B{\vee}C{\vee}D)$
- $(P{\cup}Q){\cap}(Q{\cup}R)$

Disjunctive Normal Form

A compound statement is in conjunctive normal form if it is obtained by operating OR among variables (negation of variables included) connected with ANDs. In terms of set operations, it is a compound statement obtained by Union among variables connected with Intersections.

Examples

- $(A{\wedge}B){\vee}(A{\wedge}C){\vee}(B{\wedge}C{\wedge}D)$
- $(P{\cap}Q){\cup}(Q{\cap}R)$

Proofs in Propositional Calculus

One of the main uses of a propositional calculus, when interpreted for logical applications, is to determine relations of logical equivalence between propositional formulas. These relationships are determined by means of the available transformation rules, sequences of which are called *derivations* or *proofs*.

In the discussion to follow, a proof is presented as a sequence of numbered lines, with each line consisting of a single formula followed by a *reason* or *justification* for introducing that formula. Each premise of the argument, that is, an assumption introduced as an hypothesis of the argument, is listed at the beginning of the sequence and is marked as a "premise" in lieu of other justification. The conclusion is listed on the last line. A proof is complete if every line follows from the previous ones by the correct application of a transformation rule.

Example of a Proof

- To be shown that $A \rightarrow A$.

- One possible proof of this (which, though valid, happens to contain more steps than are necessary) may be arranged as follows:

Example of a Proof	
Number	Formula
1	A
2	$A \vee A$
3	$(A \vee A) \wedge A$
4	A
5	$A \vdash A$
6	$\vdash A \rightarrow A$

Interpret $A \vdash A$ as "Assuming A, infer A". Read $\vdash A \rightarrow A$ as "Assuming nothing, infer that A implies A", or "It is a tautology that A implies A", or "It is always true that A implies A".

Soundness and completeness of the rules

The crucial properties of this set of rules are that they are *sound* and *complete*. Informally this means that the rules are correct and that no other rules are required. These claims can be made more formal as follows.

We define a *truth assignment* as a function that maps propositional variables to true or false. Informally such a truth assignment can be understood as the description of a possible state of affairs (or possible world) where certain statements are true and others are not. The semantics of formulas can then be formalized by defining for which "state of affairs" they are considered to be true, which is what is done by the following definition.

We define when such a truth assignment A satisfies a certain well-formed formula with the following rules:

- A satisfies the propositional variable P if and only if $A(P) =$ true

- A satisfies $\neg\varphi$ if and only if A does not satisfy φ

- A satisfies $(\varphi \wedge \psi)$ if and only if A satisfies both φ and ψ

- A satisfies $(\varphi \vee \psi)$ if and only if A satisfies at least one of either φ or ψ

- A satisfies $(\varphi \rightarrow \psi)$ if and only if it is not the case that A satisfies φ but not ψ

- A satisfies $(\varphi \leftrightarrow \psi)$ if and only if A satisfies both φ and ψ or satisfies neither one of them

With this definition we can now formalize what it means for a formula φ to be implied by a certain set S of formulas. Informally this is true if in all worlds that are possible given the set of formulas S the formula φ also holds. This leads to the following formal definition: We say that a set S of well-formed formulas *semantically entails* (or *implies*) a certain well-formed formula φ if all truth assignments that satisfy all the formulas in S also satisfy φ.

Finally we define *syntactical entailment* such that φ is syntactically entailed by S if and only if we can derive it with the inference rules that were presented above in a finite number of steps. This allows us to formulate exactly what it means for the set of inference rules to be sound and complete:

Soundness: If the set of well-formed formulas S syntactically entails the well-formed formula φ then S semantically entails φ.

Completeness: If the set of well-formed formulas S semantically entails the well-formed formula φ then S syntactically entails φ.

For the above set of rules this is indeed the case.

Sketch of a Soundness Proof

(For most logical systems, this is the comparatively "simple" direction of proof)

Notational conventions: Let G be a variable ranging over sets of sentences. Let A, B and C range over sentences. For "G syntactically entails A" we write "G proves A". For "G semantically entails A" we write "G implies A".

We want to show: $(A)(G)$ (if G proves A, then G implies A).

We note that "G proves A" has an inductive definition, and that gives us the immediate resources for demonstrating claims of the form "If G proves A, then ...". So our proof proceeds by induction.

    I.   Basis. Show: If A is a member of G, then G implies A.

   II.   Basis. Show: If A is an axiom, then G implies A.

 III.   Inductive step (induction on n, the length of the proof):

         a.   Assume for arbitrary G and A that if G proves A in n or fewer steps, then G implies A.

b.   For each possible application of a rule of inference at step $n + 1$, leading to a new theorem B, show that G implies B.

Notice that Basis Step II can be omitted for natural deduction systems because they have no axioms. When used, Step II involves showing that each of the axioms is a (semantic) logical truth.

The Basis steps demonstrate that the simplest provable sentences from G are also implied by G, for any G. (The proof is simple, since the semantic fact that a set implies any of its members, is also trivial.) The Inductive step will systematically cover all the further sentences that might be provable—by considering each case where we might reach a logical conclusion using an inference rule—and shows that if a new sentence is provable, it is also logically implied. (For example, we might have a rule telling us that from "A" we can derive "A or B". In III.a We assume that if A is provable it is implied. We also know that if A is provable then "A or B" is provable. We have to show that then "A or B" too is implied. We do so by appeal to the semantic definition and the assumption we just made. A is provable from G, we assume. So it is also implied by G. So any semantic valuation making all of G true makes A true. But any valuation making A true makes "A or B" true, by the defined semantics for "or". So any valuation which makes all of G true makes "A or B" true. So "A or B" is implied.) Generally, the Inductive step will consist of a lengthy but simple case-by-case analysis of all the rules of inference, showing that each "preserves" semantic implication.

By the definition of provability, there are no sentences provable other than by being a member of G, an axiom, or following by a rule; so if all of those are semantically implied, the deduction calculus is sound.

Sketch of Completeness Proof

(This is usually the much harder direction of proof.)

We adopt the same notational conventions as above.

We want to show: If G implies A, then G proves A. We proceed by contraposition: We show instead that if G does not prove A then G does not imply A. If we show that there is a model where A does not hold despite G being true, then obviously G does not imply A. The idea is to build such a model out of our very assumption that G does not prove A.

I. G does not prove A. (Assumption)

II.   If G does not prove A, then we can construct an (infinite) Maximal Set, G^*, which is a superset of G and which also does not prove A.

a.   Place an ordering (with order type ω) on all the sentences in the language (e.g., shortest first, and equally long ones in extended alphabetical ordering), and number them (E_1, E_2, ...)

b.   Define a series G_n of sets (G_0, G_1, ...) inductively:

i. $G_0 = G$

ii.   If $G_k \cup \{E_{k+1}\}$ proves A, then $G_{k+1} = G_k$

iii. If $G_k \cup \{E_{k+1}\}$ does not prove A, then $G_{k+1} = G_k \cup \{E_{k+1}\}$

c. Define G^* as the union of all the G_n. (That is, G^* is the set of all the sentences that are in any G_n.)

d. It can be easily shown that

i. G^* contains (is a superset of) G (by (b.i));

ii. G^* does not prove A (because the proof would contain only finitely many sentences and when the last of them is introduced in some G_n, that G_n would prove A contrary to the definition of G_n); and

iii. G^* is a Maximal Set with respect to A: If any more sentences whatever were added to G^*, it would prove A. (Because if it were possible to add any more sentences, they should have been added when they were encountered during the construction of the G_n, again by definition)

III. If G^* is a Maximal Set with respect to A, then it is truth-like. This means that it contains C only if it does not contain $\neg C$; If it contains C and contains "If C then B" then it also contains B; and so forth.

IV. If G^* is truth-like there is a G^*-Canonical valuation of the language: one that makes every sentence in G^* true and everything outside G^* false while still obeying the laws of semantic composition in the language.

V. A G^*-canonical valuation will make our original set G all true, and make A false.

VI. If there is a valuation on which G are true and A is false, then G does not (semantically) imply A.

Another Outline for a Completeness Proof

If a formula is a tautology, then there is a truth table for it which shows that each valuation yields the value true for the formula. Consider such a valuation. By mathematical induction on the length of the subformulas, show that the truth or falsity of the subformula follows from the truth or falsity (as appropriate for the valuation) of each propositional variable in the subformula. Then combine the lines of the truth table together two at a time by using "(P is true implies S) implies ((P is false implies S) implies S)". Keep repeating this until all dependencies on propositional variables have been eliminated. The result is that we have proved the given tautology. Since every tautology is provable, the logic is complete.

Interpretation of a Truth-functional Propositional Calculus

An interpretation of a truth-functional propositional calculus $\mathcal{P}$ is an assignment to each propositional symbol of $\mathcal{P}$ of one or the other (but not both) of the truth values truth (T) and falsity (F), and an assignment to the connective symbols of $\mathcal{P}$ of their usual truth-functional meanings. An interpretation of a truth-functional propositional calculus may also be expressed in terms of truth tables.

For n distinct propositional symbols there are 2^n distinct possible interpretations. For any particular symbol a, for example, there are $2^1 = 2$ possible interpretations:

1. a is assigned T, or

2. a is assigned F.

For the pair a, b there are $2^2 = 4$ possible interpretations:

1. both are assigned T,

2. both are assigned F,

3. a is assigned T and b is assigned F, or

4. a is assigned F and b is assigned T.

Since $\mathcal{P}$ has $\aleph_0$, that is, denumerably many propositional symbols, there are $2^{\aleph_0} = \mathfrak{c}$, and therefore uncountably many distinct possible interpretations of $\mathcal{P}$.

Interpretation of a Sentence of Truth-functional Propositional Logic

If φ and ψ are formulas of $\mathcal{P}$ and $\mathcal{I}$ is an interpretation of $\mathcal{P}$ then:

*   A sentence of propositional logic is *true under an interpretation* $\mathcal{I}$ iff $\mathcal{I}$ assigns the truth value T to that sentence. If a sentence is true under an interpretation, then that interpretation is called a *model* of that sentence.

* φ is *false under an interpretation* $\mathcal{I}$ iff φ is not true under $\mathcal{I}$.

*   A sentence of propositional logic is *logically valid* if it is true under every interpretation $\vDash \phi$ means that φ is logically valid.

*   A sentence ψ of propositional logic is a *semantic consequence* of a sentence φ iff there is no interpretation under which φ is true and ψ is false.

*   A sentence of propositional logic is *consistent* iff it is true under at least one interpretation. It is inconsistent if it is not consistent.

Some consequences of these definitions:

*   For any given interpretation a given formula is either true or false.

*   No formula is both true and false under the same interpretation.

* φ is false for a given interpretation iff $\neg\phi$ is true for that interpretation; and φ is true under an interpretation iff $\neg\phi$ is false under that interpretation.

*   If φ and $(\phi \to \psi)$ are both true under a given interpretation, then ψ is true under that interpretation.

*   If $\vDash_P \phi$ and $\vDash_P (\phi \to \psi)$, then $\vDash_P \psi$.

* $\neg\phi$ is true under $\mathcal{I}$ iff φ is not true under $\mathcal{I}$.

- $(\phi \to \psi)$ is true under $\mathcal{I}$ iff either φ is not true under $\mathcal{I}$ or ψ is true under $\mathcal{I}$.

- A sentence ψ of propositional logic is a semantic consequence of a sentence φ iff $(\phi \to \psi)$ is logically valid, that is, $\phi \vDash_p \psi$ iff $\vDash_p (\phi \to \psi)$.

Alternative Calculus

It is possible to define another version of propositional calculus, which defines most of the syntax of the logical operators by means of axioms, and which uses only one inference rule.

Axioms

Let φ, χ, and ψ stand for well-formed formulas. (The well-formed formulas themselves would not contain any Greek letters, but only capital Roman letters, connective operators, and parentheses.) Then the axioms are as follows:

Axioms		
Name	Axiom Schema	Description
THEN-1	$\phi \to (\chi \to \phi)$	Add hypothesis χ, implication introduction
THEN-2	$(\phi \to (\chi \to \psi)) \to ((\phi \to \chi) \to (\phi \to \psi))$	Distribute hypothesis φ over implication
AND-1	$\phi \wedge \chi \to \phi$	Eliminate conjunction
AND-2	$\phi \wedge \chi \to \chi$	
AND-3	$\phi \to (\chi \to (\phi \wedge \chi))$	Introduce conjunction
OR-1	$\phi \to \phi \vee \chi$	Introduce disjunction
OR-2	$\chi \to \phi \vee \chi$	
OR-3	$(\phi \to \psi) \to ((\chi \to \psi) \to (\phi \vee \chi \to \psi))$	Eliminate disjunction
NOT-1	$(\phi \to \chi) \to ((\phi \to \neg\chi) \to \neg\phi)$	Introduce negation
NOT-2	$\phi \to (\neg\phi \to \chi)$	Eliminate negation
NOT-3	$\phi \vee \neg\phi$	Excluded middle, classical logic
IFF-1	$(\phi \leftrightarrow \chi) \to (\phi \to \chi)$	Eliminate equivalence
IFF-2	$(\phi \leftrightarrow \chi) \to (\chi \to \phi)$	
IFF-3	$(\phi \to \chi) \to ((\chi \to \phi) \to (\phi \leftrightarrow \chi))$	Introduce equivalence

- Axiom THEN-2 may be considered to be a "distributive property of implication with respect to implication."

- Axioms AND-1 and AND-2 correspond to "conjunction elimination". The relation between AND-1 and AND-2 reflects the commutativity of the conjunction operator.

- Axiom AND-3 corresponds to "conjunction introduction."

- Axioms OR-1 and OR-2 correspond to "disjunction introduction." The relation between OR-1 and OR-2 reflects the commutativity of the disjunction operator.

- Axiom NOT-1 corresponds to "reductio ad absurdum."

- Axiom NOT-2 says that "anything can be deduced from a contradiction."

- Axiom NOT-3 is called "tertium non datur" (Latin: "a third is not given") and reflects the semantic valuation of propositional formulas: a formula can have a truth-value of either true or false. There is no third truth-value, at least not in classical logic. Intuitionistic logicians do not accept the axiom NOT-3.

Inference Rule

The inference rule is modus ponens:

$$\phi, \phi \to \chi \vdash \chi .$$

Meta-inference Rule

Let a demonstration be represented by a sequence, with hypotheses to the left of the turnstile and the conclusion to the right of the turnstile. Then the deduction theorem can be stated as follows:

If the sequence

$$\phi_1, \phi_2, ..., \phi_n, \chi \vdash \psi$$

has been demonstrated, then it is also possible to demonstrate the sequence

$$\phi_1, \phi_2, ..., \phi_n \vdash \chi \to \psi .$$

This deduction theorem (DT) is not itself formulated with propositional calculus: it is not a theorem of propositional calculus, but a theorem about propositional calculus. In this sense, it is a meta-theorem, comparable to theorems about the soundness or completeness of propositional calculus.

On the other hand, DT is so useful for simplifying the syntactical proof process that it can be considered and used as another inference rule, accompanying modus ponens. In this sense, DT corresponds to the natural conditional proof inference rule which is part of the first version of propositional calculus introduced in this topic.

The converse of DT is also valid:

If the sequence

$$\phi_1, \phi_2, ..., \phi_n \vdash \chi \to \psi$$

has been demonstrated, then it is also possible to demonstrate the sequence

$$\phi_1, \phi_2, ..., \phi_n, \chi \vdash \psi$$

in fact, the validity of the converse of DT is almost trivial compared to that of DT:

If

$$\phi_1, ..., \phi_n \vdash \chi \rightarrow \psi$$

then

1: $\phi_1, ..., \phi_n, \chi \vdash \chi \rightarrow \psi$

2: $\phi_1, ..., \phi_n, \chi \vdash \chi$

and from (1) and (2) can be deduced

3: $\phi_1, ..., \phi_n, \chi \vdash \psi$

by means of modus ponens, Q.E.D.

The converse of DT has powerful implications: it can be used to convert an axiom into an inference rule. For example, the axiom AND-1,

$$\vdash \phi \wedge \chi \rightarrow \phi$$

can be transformed by means of the converse of the deduction theorem into the inference rule

$$\phi \wedge \chi \vdash \phi$$

which is conjunction elimination, one of the ten inference rules used in the first version of the propositional calculus.

Example of a Proof

The following is an example of a (syntactical) demonstration, involving only axioms THEN-1 and THEN-2:

Prove: $A \rightarrow A$ (Reflexivity of implication).

Proof:

1. $(A \rightarrow ((B \rightarrow A) \rightarrow A)) \rightarrow ((A \rightarrow (B \rightarrow A)) \rightarrow (A \rightarrow A))$

   Axiom THEN-2 with $\phi = A, \chi = B \rightarrow A, \psi = A$

2. $A \rightarrow ((B \rightarrow A) \rightarrow A)$

   Axiom THEN-1 with $\phi = A, \chi = B \rightarrow A$

3. $(A \rightarrow (B \rightarrow A)) \rightarrow (A \rightarrow A)$

   From (1) and (2) by modus ponens.

4. $A \rightarrow (B \rightarrow A)$

   Axiom THEN-1 with $\phi = A, \chi = B$

5. $A \to A$

From (3) and (4) by modus ponens.

Equivalence to Equational Logics

The preceding alternative calculus is an example of a Hilbert-style deduction system. In the case of propositional systems the axioms are terms built with logical connectives and the only inference rule is modus ponens. Equational logic as standardly used informally in high school algebra is a different kind of calculus from Hilbert systems. Its theorems are equations and its inference rules express the properties of equality, namely that it is a congruence on terms that admits substitution.

Classical propositional calculus as described above is equivalent to Boolean algebra, while intuitionistic propositional calculus is equivalent to Heyting algebra. The equivalence is shown by translation in each direction of the theorems of the respective systems. Theorems ϕ of classical or intuitionistic propositional calculus are translated as equations $\phi = 1$ of Boolean or Heyting algebra respectively. Conversely theorems $x = y$ of Boolean or Heyting algebra are translated as theorems $(x \to y) \wedge (y \to x)$ of classical or intuitionistic calculus respectively, for which $x \equiv y$ is a standard abbreviation. In the case of Boolean algebra $x = y$ can also be translated as $(x \wedge y) \vee (\neg x \wedge \neg y)$, but this translation is incorrect intuitionistically.

In both Boolean and Heyting algebra, inequality $x \leq y$ can be used in place of equality. The equality $x = y$ is expressible as a pair of inequalities $x \leq y$ and $y \leq x$. Conversely the inequality $x \leq y$ is expressible as the equality $x \wedge y = x$, or as $x \vee y = y$. The significance of inequality for Hilbert-style systems is that it corresponds to the latter's deduction or entailment symbol $\vdash$. An entailment

$$\phi_1, \phi_2, \ldots, \phi_n \vdash \psi$$

is translated in the inequality version of the algebraic framework as

$$\phi_1 \wedge \phi_2 \wedge \ldots \wedge \phi_n \leq \psi$$

Conversely the algebraic inequality $x \leq y$ is translated as the entailment

$$x \vdash y .$$

The difference between implication $x \to y$ and inequality or entailment $x \leq y$ or $x \vdash y$ is that the former is internal to the logic while the latter is external. Internal implication between two terms is another term of the same kind. Entailment as external implication between two terms expresses a metatruth outside the language of the logic, and is considered part of the metalanguage. Even when the logic under study is intuitionistic, entailment is ordinarily understood classically as two-valued: either the left side entails, or is less-or-equal to, the right side, or it is not.

Similar but more complex translations to and from algebraic logics are possible for natural deduction systems as described above and for the sequent calculus. The entailments of the latter can be interpreted as two-valued, but a more insightful interpretation is as a set, the elements of which can be understood as abstract proofs organized as the morphisms of a category. In this interpretation the cut rule of the sequent calculus corresponds to composition in the category. Boolean and Heyting algebras enter this picture as special categories having at most one morphism per homset, i.e., one proof per entailment, corresponding to the idea that existence of proofs is all that matters: any proof will do and there is no point in distinguishing them.

Graphical Calculi

It is possible to generalize the definition of a formal language from a set of finite sequences over a finite basis to include many other sets of mathematical structures, so long as they are built up by finitary means from finite materials. What's more, many of these families of formal structures are especially well-suited for use in logic.

For example, there are many families of graphs that are close enough analogues of formal languages that the concept of a calculus is quite easily and naturally extended to them. Indeed, many species of graphs arise as *parse graphs* in the syntactic analysis of the corresponding families of text structures. The exigencies of practical computation on formal languages frequently demand that text strings be converted into pointer structure renditions of parse graphs, simply as a matter of checking whether strings are well-formed formulas or not. Once this is done, there are many advantages to be gained from developing the graphical analogue of the calculus on strings. The mapping from strings to parse graphs is called *parsing* and the inverse mapping from parse graphs to strings is achieved by an operation that is called *traversing* the graph.

Other Logical Calculi

Propositional calculus is about the simplest kind of logical calculus in current use. It can be extended in several ways. (Aristotelian "syllogistic" calculus, which is largely supplanted in modern logic, is in *some* ways simpler – but in other ways more complex – than propositional calculus.) The most immediate way to develop a more complex logical calculus is to introduce rules that are sensitive to more fine-grained details of the sentences being used.

First-order logic (a.k.a. first-order predicate logic) results when the "atomic sentences" of propositional logic are broken up into terms, variables, predicates, and quantifiers, all keeping the rules of propositional logic with some new ones introduced. (For example, from "All dogs are mammals" we may infer "If Rover is a dog then Rover is a mammal".) With the tools of first-order logic it is possible to formulate a number of theories, either with explicit axioms or by rules of inference, that can themselves be treated as logical calculi. Arithmetic is the best known of these; others include set theory and mereology. Second-order logic and other higher-order logics are formal extensions of first-order logic. Thus, it makes sense to refer to propositional logic as *"zeroth-order logic"*, when comparing it with these logics.

Modal logic also offers a variety of inferences that cannot be captured in propositional calculus. For example, from "Necessarily p" we may infer that p. From p we may infer "It is possible that p". The translation between modal logics and algebraic logics concerns classical and intuitionistic logics but with the introduction of a unary operator on Boolean or Heyting algebras, different from the Boolean operations, interpreting the possibility modality, and in the case of Heyting algebra a second operator interpreting necessity (for Boolean algebra this is redundant since necessity is the De Morgan dual of possibility). The first operator preserves 0 and disjunction while the second preserves 1 and conjunction.

Many-valued logics are those allowing sentences to have values other than *true* and *false*. (For example, *neither* and *both* are standard "extra values"; "continuum logic" allows each sentence to have any of an infinite number of "degrees of truth" between *true* and *false*.) These logics often require calculational devices quite distinct from propositional calculus. When the values form a Boolean algebra (which may have more than two or even infinitely many values), many-valued logic reduces to classical logic; many-valued logics are therefore only of independent interest when the values form an algebra that is not Boolean.

Categorical Logic

Categorical logic is a relatively new field arising from the application of the mathematical theory of categories to logic and theoretical computer science. Category theory consists of a characteristic language and collection of methods and results that have become common-place in many mathematics-based disciplines. It is a branch of abstract algebra invented in the tradition of Felix Klein's Erlanger Programm as a way of studying different kinds of mathematical structures in terms of their "admissible transformations". The general notion of a category provides an axiomatization of the notion of a "structure-preserving transformation", and thereby of a species of structure admitting such transformations. As an abstract theory of mappings, with such great generality, it is not surprising that category theory should have wide-spread applications in many types of foundational work.

The applications of category theory in logic often involve the use of topology, sheaf theory, and other ideas imported from geometry, particularly in constructing models. This occurs, for example, in domain theory or topos theory. But as in algebraic topology, where category theory was first invented, extensive use is also made of algebraic techniques, for example in the treatment of logical theories as "generalized algebras". In this way, categorical logic typically treats the classical, logical notions of semantics as "geometry" and syntax as a kind of "algebra", to which general category theory can then be applied, in order to study the connections between the two.

Categories as Sets

We assume there is some domain of discourse, a universal set or universe S. (This corresponds to outcome space in a random experiment.) The universe contains all the "things" that can belong to categories. A category is any subset of the universe.

Suppose A and B are sets. We will be concerned with statements like "every element of A is an element of B," which we might write as "every A is a B" or "all A are B." Of course, that is just what it means for A to be a subset of B: $A \subset B$. An example: If A comprises all ravens and B comprises all birds, then $A \subset B$ could be pronounced everything that is a raven is a bird, every raven is a bird, or all ravens are birds.

Another categorical statement is the denial of "all A are B ," namely, "some A are not B." That is just what it means for A not to be a subset of B : $A \not\subset B$. An example is some birds are not ravens. Note that some birds are not ravens is not equivalent to some ravens are not birds: $A \not\subset B$ is not the same as $B \not\subset A$.

Interestingly, the two statements, A⊂B and A⊄B, have a fundamental difference: it can be true that A⊂B even if A has no elements and B has no elements, because the empty set is a subset of every set. For example, it is true that all immortal ravens are pink birds, because there are no immortal ravens.

In contrast, if $A \not\subset B$, A must have elements, at least one of which is in B^c. For instance, it is not true that some immortal ravens are not pink birds, because that would require there to be some immortal ravens. Hence $A \not\subset B$ lets us deduce that neither A nor B^c is empty: the relationship ⊄ has existential import while the relationship ⊂ does not.

Be careful not to assume that any set has a member unless the diagram represents "some are" or "some are not." To assume without justification that a set has at least one element (i.e., that an element of the set exists) is called the existential fallacy.

There are two other categorical statements, "some A are B." and "no A are B." The first can be rephrased as "some A are not non-B," that is, $A \not\subset B^c$. An example would be some birds are ravens; equivalently, not every bird is a non-raven, or (even more geeky) birds are not a subset of non-ravens. The second can be rephrased as "all A are non-B," that is, $A \subset B^c$. An example might be no invisible birds are ravens or every invisible bird is a non-raven.

Later in this chapter we will look at other ways of expressing these four categorical relationships: (i) all A are B, (ii) some A are not B, (iii) some A are B, (iv) all A are not B. As just illustrated, the statements with the word some imply that something (a member of A, at least) exists—they have existential import. The statements with all do not imply that A has any elements: they have no existential import.

For example, in the definition A={all animals that are birds}, the thing that is true for the members of the set is that they are birds. That is, the name of the category of things, "birds" is essentially the membership function: the function that is true for birds and false for everything else.

For instance, if S is the set of all numbers, we could write the set of even numbers as $\{x \in S : \frac{x}{2}$ is an integer$\}$. The colon is pronounced "such that." This expression means "the set of all numbers x such that when you divide x by 2, the result is an integer. The membership function here is $x/2$ is an integer, which is true if x is an even number and false otherwise. In general, to specify a set by its membership function, we use an expression like $\{x \in S :$ (the membership function is true for x)$\}$.

The membership function of a set A is like an oracle: you bring it an element x of S and the membership function says, "yes, this is an element of A," or "no, this is not an element of A." It assigns the value "true" or the value "false" to elements of S, according to whether they are elements of A. If $P()$ is the membership function for the set A, then $A = \{x \in S : P(x)\}$. When S is clear from context, we can omit it from the notation and just write $A = \{x : P(x)\}$.

A set is equivalent to its membership function: if you know one, you know the other. Because of that equivalence, we will often use the same symbol to denote a set and its membership function. For instance, unadorned, A might denote a set of elements of S, while with parentheses, $A()$ might denote a function that assigns the value "true" to elements of A and the value "false" to all other elements of $S : A = \{x : A(x)\}$.

To bring the discussion back to Earth, suppose the domain of discourse S is birds. Black birds comprise a subset of S. Let B be the class of black birds:

$B=\{x \in S : x$ is black$\}$.

Let R denote the set of ravens, a different class of birds.

$R=\{x \in S : x$ is a raven$\}$.

Then we might ask whether the following argument is valid:

Some ravens are black. All ravens are birds. Therefore, some birds are black.

Before answering the question, we will introduce some standard mathematical notation.

Existential and Universal Quantifiers

We shall be concerned with statements like "every element of A is an element of B " (informally, "all A are B," "every A is a B"), "some element of B is not an element of A " (informally, "some B are not A," "not all B are A," "not every B is an A"), and the like. As discussed more below, *every* and *all* do not presuppose that there are any. In contrast, *some* means at least one.

The symbol $\forall$ $\forall$ means every or all. It is called the universal quantifier. For instance, the statement all ravens are black (birds) could be symbolized as:

$\forall x \in S$, *if* $x \in R$ *then* $x \in B$.

We could also write simply $R \subset B$. Here is the corresponding Venn diagram:

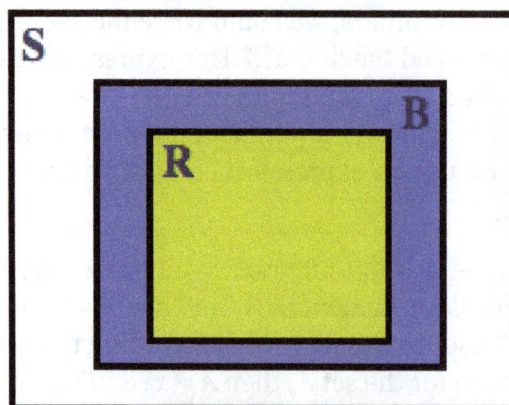

Venn Diagram for All R are B

Note that all ravens are black birds is quite different from all black birds are ravens: $R \subset B$ is not the same as $B \subset R$.

The statement no ravens are black birds could be symbolized as:

$x \in S$, if $x \in R$ then $x \notin B$. $x \in S$, if $x \in R$ then $x \notin B$.

We could also write simply R∩B={} R∩B={}; that is, RR and B B are disjoint. Here is the corresponding Venn diagram:

Venn Diagram for No R is B

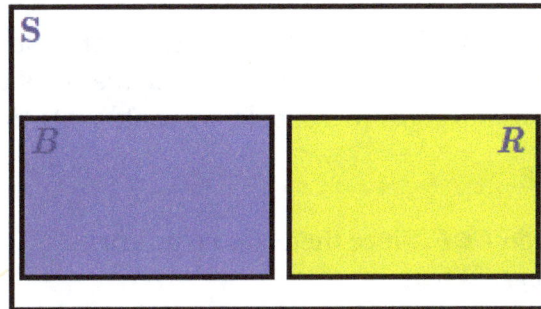

Here, in contrast, *no ravens are black birds* describes the same state of the world as *no black birds are ravens*: R∩B={} f and only if B∩R={}.

The symbol ∃ means *there exists, some* or *at least one*. It is called the *existential quantifier. Some* means at least one, but not necessarily all. The statement *some raven is black* could be symbolized as:

$\exists x \in S : x \in R$ and $x \in B$.

We could write R∩B≠{} (something is both in R and in B: there is a black raven), $R \not\subset B^c$ (not every element of R is not in B: not every raven is non-black), or $B \not\subset R^c$ (not every element of B is not in R: not every black thing is a non-raven). Here is a Venn diagram illustrating the situation.

Venn Diagram for Some R is B.

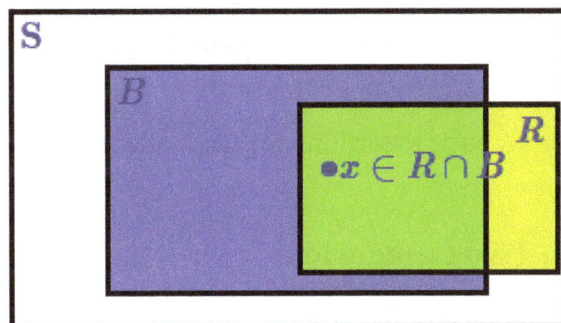

The statement some ravens are black birds describes the same situation as some black birds are ravens: $R \not\subset B^c$ if and only if $B \not\subset R^c$. Both conditions are true exactly when the intersection of R and B s at least one element.

Let Y={x∈S : x is yellow} be the set of yellow birds. The assertion that no ravens are yellow could be written R∩Y={} (no bird is both a raven and yellow), $R \subset Y^c$ (every raven is non-yellow), or $Y \subset R^c$ (every yellow bird is a non-raven).

Naively, we might assume that if something is true for *every* element of some set, it is true for at least *one* element of the set. But for the fact that the set could be empty, that is true. However, if the set in question is empty, something can be true for all of its elements and yet not true for at least one of its elements. This is an important difference between ∃ and ∀: the former asserts that something exists, while the latter can be satisfied even if nothing exists.

Here is an example. Let the universe be the set S of animals. Let us suppose that unicorns do not exist. Let U be the set of animals that are unicorns. Let E be the set of animals that lay eggs. Consider the assertion *All unicorns lay eggs*. In symbols, we could write that as

　　　$\forall x \in S$, if $x \in U$ then $x \in E$.

Equivalently, it is the assertion $U \subset E$. Since there are no unicorns, U is the empty set, and the assertion that all unicorns lay eggs is true.

Now consider the assertion *Some unicorns lay eggs*. In symbols, we could write that as

　　　$\exists x \in S : x \in U$ and $x \in E$.

Equivalently, it is the assertion $U \cap E \neq \{\}$ or $U \not\subset E^c$. Since there are no unicorns, U is the empty set, which is a subset of every set, including E^c; similarly $U \cap E$ is also the empty set, and the assertion that some unicorns lay eggs is false.

The assertion *no unicorns lay eggs* can be written $\forall x \in S$, if $x \in U$ then $x \notin E$, or $U \subset E^c$. Since there are no unicorns, this assertion is true. The assertion *some unicorns do not lay eggs* can be written $\exists x \in S : x \in U$ and $x \notin E$, $U \cap E^c \neq \{\}$, or $U \not\subset E$ This assertion is false—it must be because U is empty.

Here is a more mathematical explanation. Suppose that the set A is empty. Consider the assertion $\forall x \in S$, if $x \in A$ then $x \in B$. That is equivalent to the assertion that $A \subset B$, which is true because the empty set $\{\}$ is a subset of every set, including, in particular, B.

Now consider the assertion $\exists x \in S : x \in A$ and $x \in B$. That is equivalent to the assertion that $A \cap B \neq \{\}$ and to the assertion $A \not\subset B^c$. The assertion is false if the set A is empty. It is true that every element of the set A is an element of the set B, and yet there is nothing that is an element of the set A and also an element of the set B.

In categorical logic, there are six basic assertions we can make regarding the relation between two classes, listed below.

Relationships between Two Categories

Suppose we have two categories, A and B, subsets of a domain of discourse S. We can express six categorical relations between A and B. The relations can be written using universal and existential quantifiers, or using set notation:

- Every A is a B; All A are B.

　　$\forall x$, if $x \in A$ then $x \in B$.

　　$A \subset B$.

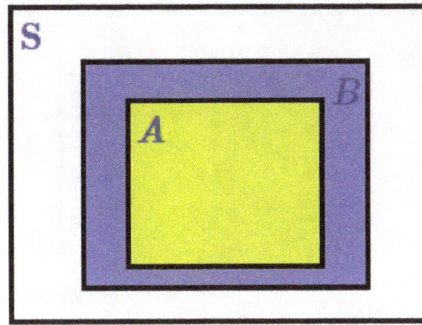

- Every B is an A; All B are A .

  $\forall x$, if $x \in B$ then $x \in A$.

  $B \subset A$.

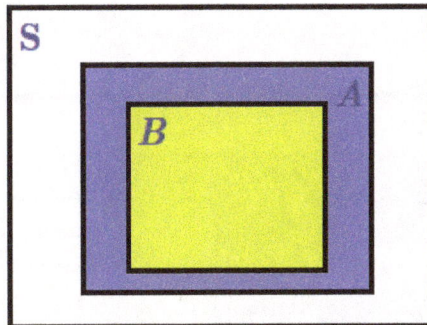

- No A is a B; No B is an A.

  $\forall x$, if $x \in A$ then $x \notin B$.

  $A \subset B^c$

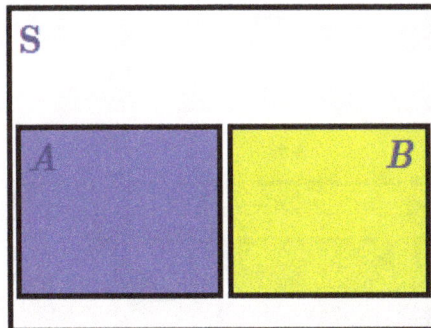

- Some A is a B; Some B is an.

  $\exists x : x \in A$ and $x \in B$.

  $A \cap B \neq \{\}$.

  $A \not\subset B^c$.

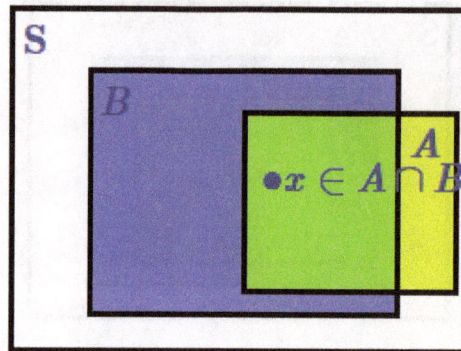

- Some A is not a B; Not all A are B.

  $\exists x : x \in A$ and $x \notin B$.

  $A \cap B^c \neq \{\}$.

  $A \not\subset B$.

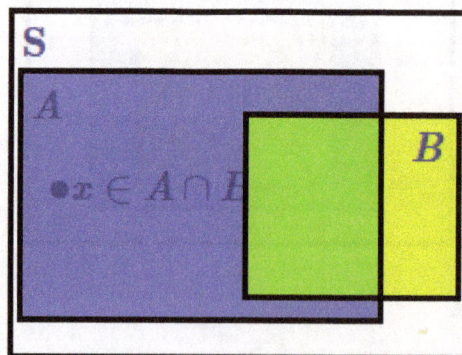

- Some B is not an A; Not all B are A.

  $\exists x : x \in B$ and $x \notin A$.

  $B \cap A^c \neq \{\}$.

  $B \not\subset A$.

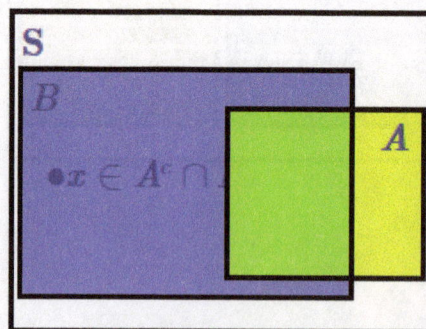

As you can see, set notation is more compact.

There is more than one way to express any quantified assertion. For example, some A is not a B can also be written not all A are B, A∩B^c ≠{} and A⊄B. As you read this chapter, draw Venn diagrams to represent each assertion. Use the Venn diagram to help you find other, equivalent assertions. If the diagram involves A and B, think about the relationship between A^c and B, between A and B^c, and between A^c and B^c.

For instance, here is the Venn diagram for every A is a B:

Venn Diagram for All A are B

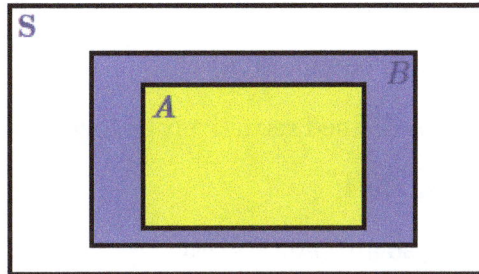

In the diagram, the area that represents *A* is entirely contained in the area that represents *B*. The area that represents the complement of *B* is entirely contained in the area that represents the complement of *A*. No part of *A* is in the complement of *B*. The relationship *every A is a B* is mathematically equivalent to the relationship *every non-B is a non-A* and to the relationship *no A is a non-B*.

Similarly, here is the Venn diagram for some A are B:

Venn Diagram for Some A are B

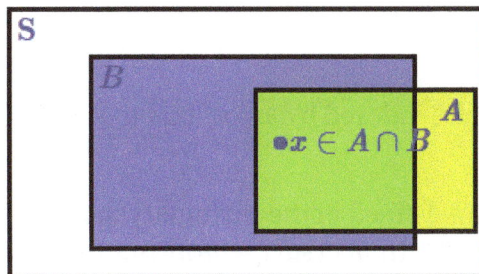

You can see from the diagram that the situation could also be described as some B are A, not every A is a non-B, and not every B is a non-A.

Infinitary Logic

An infinitary logic is a logic that allows infinitely long statements or infinitely long proofs, for example, by allowing conjunctions, disjunctions and quantifier sequences, to be of infinite length. The price to pay for adopting many of these more expressive logics is the failure of completeness or of compactness.

Given a pair κ, λ of infinite cardinals such that $\lambda \leq \kappa$, we define a class of infinitary languages in each of which we may form conjunctions and disjunctions of sets of formulas of cardinality $< \kappa$, and quantifications over sequences of variables of length $< \lambda$.

Let L — the (finitary) *base language* — be an arbitrary but fixed first-order language with any number of extralogical symbols. The infinitary language $L(\kappa,\lambda)$ has the following *basic symbols*:

- All symbols of L

- A set Var of individual variables, where the cardinality of Var (written: |Var|) is κ

- A logical operator $\wedge$ (*infinitary conjunction*)

The class of *preformulas* of $L(\kappa,\lambda)$ is defined recursively as follows:

- Each formula of L is a preformula;

- If φ and ψ are preformulas, so are $\varphi \wedge \psi$ and $\neg\varphi$;

- If Φ is a set of preformulas such that $|\Phi| < \kappa$, then $\wedge\Phi$ is a preformula;

- If φ is a preformula and $X \subseteq$ Var is such that $|X| < \lambda$, then $\exists X\varphi$ is a preformula;

- All preformulas are defined by the above clauses.

If Φ is a set of preformulas indexed by a set I, say $\Phi = \{\varphi_i : i \in I\}$, then we agree to write$\wedge\Phi$ for:

$$\wedge_{i\in I}\varphi$$

or, if I is the set of natural numbers, we write $\wedge\Phi$ for:

$$\varphi_0 \wedge \varphi_1 \wedge \dots$$

If X is a set of individual variables indexed by an ordinal α, say $X = \{x_\xi : \xi < \alpha\}$, we agree to write $(\exists x_\xi)_{\xi<\alpha}\varphi$ for $\exists X\varphi$.

The logical operators $\vee$, $\rightarrow$, $\leftrightarrow$ are defined in the customary manner. We also introduce the operators $\vee$ (*infinitary disjunction*) and $\forall$ (*universal quantification*) by

$$\vee\Phi =_{df} \neg\wedge\{\neg\varphi : \varphi \in \Phi\}$$

$$\forall X\varphi =_{df} \neg\exists X\neg\varphi,$$

and employ similar conventions as for $\wedge$, $\exists$.

Thus $L(\kappa,\lambda)$ is the infinitary language obtained from L by permitting conjunctions and disjunctions of length$< \kappa$ and quantifications of length $< \lambda$. Languages $L(\kappa,\omega)$ are called *finite-quantifier* languages, the rest *infinite-quantifier* languages. Observe that $L(\omega,\omega)$ is just L itself.

Notice the following *anomaly* which can arise in an infinitary language but not in a finitary one. In the language $L(\omega_1,\omega)$, which allows countably infinite conjunctions but only finite quantifications, there are preformulas with so many free variables that they cannot be "closed" into sentences of $L(\omega_1,\omega)$ by prefixing quantifiers. Such is the case, for example, for the $L(\omega_1,\omega)$-preformula:

$$x_0 < x_1 \wedge x_1 < x_2 \wedge \ldots \wedge x_n < x_{n+1} \ldots,$$

where L contains the binary relation symbol $<$. For this reason we make the following

Definition: A *formula* of $L(\kappa,\lambda)$ is a preformula which contains $< \lambda$ free variables. The set of all formulas of $L(\kappa,\lambda)$ will be denoted by Form($L(\kappa,\lambda)$) or simply Form(κ,λ) and the set of all sentences by Sent($L(\kappa,\lambda)$) or simply Sent(κ,λ).

In this connection, observe that, in general, nothing would be gained by considering "languages" $L(\kappa,\lambda)$ with $\lambda > \kappa$. For example, in the "language" $L(\omega,\omega_1)$, formulas will have only finitely many free variables, while there will be a host of "useless" quantifiers able to bind infinitely many free variables.

Having defined the syntax of $L(\kappa,\lambda)$, we next sketch its *semantics*. Since the extralogical symbols of $L(\kappa,\lambda)$ are just those of L, and it is these symbols which determine the form of the structures in which a given first-order language is to be interpreted, it is natural to define an $L(\kappa,\lambda)$-structure to be simply an L-structure. The notion of a formula of $L(\kappa,\lambda)$ being *satisfied* in an L-structure A (by a sequence of elements from the domain of A) is defined in the same inductive manner as for formulas of L except that we must add two extra clauses corresponding to the clauses for $\wedge\Phi$ and $\exists X\varphi$ in the definition of preformula. In these two cases we naturally define:

$\wedge\Phi$ is satisfied in A (by a given sequence) $\Leftrightarrow$ for all $\varphi \in \Phi$, φ is satisfied in A (by the sequence);

$\exists X\varphi$ is satisfied in A $\Leftrightarrow$ there is a sequence of elements from the domain of A in bijective correspondence with X which satisfies φ in A.

These informal definitions need to be tightened up in a rigorous development, but their meaning should be clear to the reader. Now the usual notions of *truth, validity, satisfiability,* and *model* for formulas and sentences of $L(\kappa,\lambda)$ become available. In particular, if A is an L-structure and $\sigma \in$ Sent(κ,λ), we shall write $A \vDash \sigma$ for A is a model of σ, and $\vDash \sigma$ for σ *is valid*, that is, for all A, $A \vDash \sigma$. If $\Delta \subseteq$ Sent(κ,λ), we shall write $\Delta \vDash \sigma$ for σ *is a logical consequence of* Δ, that is, each model of Δ is a model of σ.

We now give some examples intended to display the expressive power of the infinitary languages $L(\kappa,\lambda)$ with $\kappa \geq \omega_1$. In each case it is well-known that the notion in question cannot be expressed in any first-order language.

Characterization of the standard model of arithmetic in $L(\omega_1,\omega)$. Here the *standard model of arithmetic* is the structure N = $\langle N, +, \cdot, s, 0 \rangle$, where N is the set of natural numbers, $+$, $\cdot$, and 0 have their usual meanings, and s is the successor operation. Let L be the first-order language appropriate for N. Then the class of L-structures isomorphic to N coincides with the class of models of the conjunction of the following $L(\omega_1,\omega)$ sentences (where 0 is a name of 0):

$$\bigwedge_{m\in\omega} \bigwedge_{n\in\omega} s^m 0 + s^n 0 = s^{m+n} 0$$

$$\bigwedge_{m\in\omega} \bigwedge_{n\in\omega} s^m 0 \cdot s^n 0 = s^{m\cdot n} 0$$

$$\wedge_{m\in\omega} \wedge_{n\in\omega-\{m\}} s^m 0 \neq s^n 0$$

$$\forall x \vee_{m\in\omega} x = s^m 0$$

The terms $s^n x$ are defined recursively by

$$s^0 x \quad = \quad x$$

$$s^{n+1} x \quad = \quad s(s^n x)$$

Characterization of the class of all finite sets in $L(\omega_1,\omega)$. Here the base language has no extralogical symbols. The class of all finite sets then coincides with the class of models of the $L(\omega_1,\omega)$-sentence

$$\vee_{n\in\omega} \exists v_0 \ldots \exists v_n \forall x (x = v_0 \vee \ldots \vee x = v_n).$$

Truth definition in $L(\omega_1,\omega)$ for a countable base language L. Let L be a countable first-order language (for example, the language of arithmetic or set theory) which contains a name n for each natural number n, and let $\sigma_0, \sigma_1, \ldots$ be an enumeration of its sentences. Then the $L(\omega_1,\omega)$-formula

$$\text{Tr}(x) =_{df} \vee_{n\in\omega} (x = n \wedge \sigma_n)$$

is a *truth predicate* for L inasmuch as the sentence

$$\text{Tr}(n) \leftrightarrow \sigma_n$$

is valid for each n.

Characterization of well-orderings in $L(\omega_1,\omega_1)$. The base language L here includes a binary predicate symbol $\leq$. Let σ_1 be the usual L-sentence characterizing linear orderings. Then the class of L-structures in which the interpretation of $\leq$ is a well-ordering coincides with the class of models of the $L(\omega_1,\omega_1)$ sentence $\sigma = \sigma_1 \wedge \sigma_2$, where

$$\sigma_2 =_{df} (\forall v_n)_{n\in\omega} \exists x [\vee_{n\in\omega} (x = v_n) \wedge \wedge_{n\in\omega} (x \leq v_n)].$$

Notice that the sentence σ_2 contains an *infinite quantifier*: it expresses the essentially *second-order* assertion that every countable subset has a least member. It can in fact be shown that the presence of this infinite quantifier is essential: the class of well-ordered structures cannot be characterized in any finite-quantifier language. This example indicates that infinite-quantifier languages such as $L(\omega_1,\omega_1)$ behave rather like second-order languages; we shall see that they share the latters' defects (incompleteness) as well as some of their advantages (strong expressive power).

Many extensions of first-order languages can be *translated* into infinitary languages. For example, consider the generalized quantifier language $L(Q_0)$ obtained from L by adding a new quantifier symbol Q_0 and interpreting $Q_0 x\varphi(x)$ as *there exist infinitely many x such that* $\varphi(x)$. It is easily seen that the sentence $Q_0 x\varphi(x)$ has the same models as the $L(\omega_1,\omega)$-sentence

$$\neg\vee_{n\in\omega} \exists v_0 \ldots \exists v_n \forall x [\varphi(x) \rightarrow (x = v_0 \vee \ldots \vee x = v_n)].$$

Thus $L(Q_0)$ is, in a natural sense, translatable into $L(\omega_1,\omega)$. Another language translatable into $L(\omega_1,\omega)$ in this sense is the *weak second-order language* obtained by adding a countable set of monadic predicate variables to L which are then interpreted as ranging over all *finite* sets of individuals.

Languages with arbitrarily long conjunctions, disjunctions and (possibly) quantifications may also be introduced. For a fixed infinite cardinal λ, the language $L(\infty,\lambda)$ is defined by specifying its class of formulas, Form(∞,λ), to be the union, over all $\kappa \geq \lambda$, of the sets Form(κ,λ). Thus $L(\infty,\lambda)$ allows arbitrarily long conjunctions and disjunctions, in the sense that if Φ is an arbitrary subset of Form(∞,λ), then both $\wedge\Phi$ and $\vee\Phi$ are members of Form(∞,λ). But $L(\infty,\lambda)$ admits only quantifications of length $< \lambda$: all its formulas have $< \lambda$ free variables. The language $L(\infty,\infty)$ is defined in turn by specifying its class of formulas, Form(∞,∞), to be the union, over all infinite cardinals λ, of the classes Form(∞,λ). So $L(\infty,\infty)$ allows arbitrarily long quantifications in addition to arbitrarily long conjunctions and disjunctions. Note that Form(∞,λ) and Form(∞,∞) are proper classes in the sense of Gödel-Bernays set theory. Satisfaction of formulas of $L(\infty,\lambda)$ and $L(\infty,\infty)$ in a structure may be defined by an obvious extension of the corresponding notion for $L(\kappa,\lambda)$.

Hilbert-type Infinitary Logics

A first-order infinitary logic $L_{\alpha,\beta}$, α regular, $\beta = 0$ or $\omega \leq \beta \leq \alpha$, has the same set of symbols as a finitary logic and may use all the rules for formation of formulae of a finitary logic together with some additional ones:

- Given a set of formulae $A = \{A_\gamma \mid \gamma < \delta < \alpha\}$ then $(A_0 \vee A_1 \vee \cdots)$ and $(A_0 \wedge A_1 \wedge \cdots)$ are formulae. (In each case the sequence has length δ.)

- Given a set of variables $V = \{V_\gamma \mid \gamma < \delta < \beta\}$ and a formula A_0 then $\forall V_0 : \forall V_1 \cdots (A_0)$ and $\exists V_0 : \exists V_1 \cdots (A_0)$ are formulae. (In each case the sequence of quantifiers has length δ.)

The concepts of free and bound variables apply in the same manner to infinite formulae. Just as in finitary logic, a formula all of whose variables are bound is referred to as a *sentence*.

A theory T in infinitary logic $L_{\alpha,\beta}$ is a set of sentences in the logic. A proof in infinitary logic from a theory T is a sequence of statements of length γ which obeys the following conditions: Each statement is either a logical axiom, an element of T, or is deduced from previous statements using a rule of inference. As before, all rules of inference in finitary logic can be used, together with an additional one:

- Given a set of statements $A = \{A_\gamma \mid \gamma < \delta < \alpha\}$ which have occurred previously in the proof then the statement $\wedge_{\gamma<\delta} A_\gamma$ can be inferred.

The logical axiom schemata specific to infinitary logic are presented below. Global schemata variables: δ and γ such that $0 < \delta < \alpha$.

- $((\wedge_{\epsilon<\delta}(A_\delta \Rightarrow A_\epsilon)) \Rightarrow (A_\delta \Rightarrow \wedge_{\epsilon<\delta} A_\epsilon))$

- For each $\gamma < \delta, ((\wedge_{\epsilon<\delta} A_\epsilon) \Rightarrow A_\gamma)$

- Chang's distributivity laws (for each γ): $(\vee_{\mu<\gamma}(\wedge_{\delta<\gamma} A_{\mu,\delta}))$, where $\forall\mu\forall\delta\exists\epsilon < \gamma : A_{\mu,\delta} = A_\epsilon$ or $A_{\mu,\delta} = \neg A_\epsilon$, and $\forall g \in \gamma^\gamma \exists \epsilon < \gamma : \{A_\epsilon, \neg A_\epsilon\} \subseteq \{A_{\mu,g(\mu)} : \mu < \gamma\}$

- For $\gamma < \alpha$, $((\wedge_{\mu<\gamma}(\vee_{\delta<\gamma} A_{\mu,\delta})) \Rightarrow (\vee_{\epsilon<\gamma^\gamma}(\wedge_{\mu<\gamma} A_{\mu,\gamma_\epsilon(\mu)})))$, where $\{\gamma_\epsilon : \epsilon < \gamma^\gamma\}$ is a well ordering of γ^γ

The last two axiom schemata require the axiom of choice because certain sets must be well orderable. The last axiom schema is strictly speaking unnecessary as Chang's distributivity laws imply it, however it is included as a natural way to allow natural weakenings to the logic.

Completeness, Compactness and Strong Completeness

A theory is any set of statements. The truth of statements in models are defined by recursion and will agree with the definition for finitary logic where both are defined. Given a theory T a statement is said to be valid for the theory T if it is true in all models of T.

A logic $L_{\alpha,\beta}$ is complete if for every sentence S valid in every model there exists a proof of S. It is strongly complete if for any theory T for every sentence S valid in T there is a proof of S from T. An infinitary logic can be complete without being strongly complete.

A cardinal $\kappa \neq \omega$ is weakly compact when for every theory T in $L_{\kappa,\kappa}$ containing at most κ any formulas, if every $S \subseteq T$ of cardinality less than κ has a model, then T has a model. A cardinal $\kappa \neq \omega$ is strongly compact when for every theory T in $L_{\kappa,\kappa}$, without restriction on size, if every $S \subseteq T$ of cardinality less than κ has a model, then T has a model.

Algebraic Logic

The branch of mathematical logic that deals with propositions from the aspect of their logical meanings (true or false) and with logical operations on them.

The algebra of logic originated in the middle of the 19th century with the studies of G. Boole, and was subsequently developed by C.S. Peirce, P.S. Poretskii, B. Russell, D. Hilbert, and others. The development of the algebra of logic was an attempt to solve traditional logical problems by algebraic methods. With the birth of the theory of sets (in the 1870s), propositions and logical operations on them became the principal subject of algebra of logic. Propositions are understood to mean statements about which it is meaningful to ask whether they are true or false. For instance, the proposition "a whale is an animal" is true, while the statement "all angles are right angles" is false. The connectives "and" , "or" , "if ... then" , "is equivalent to" , the particle "not" , etc., which are commonly used in the language of logic, make it possible to construct new, more "complicated" , propositions from those already established. Thus, given that "x> 2" and "x≤ 3" , it is possible to obtain, by using the connective "and" , the proposition "x> 2 and x≤ 3" ; by using the connective "or" it is possible to obtain the proposition "x> 2 or x≤ 3" , etc.

The truth or the falsehood of the propositions obtained in this way depend on the truth or falsehood of the initial propositions and on a corresponding treatment of the connectives as operations on propositions. Often truth is symbolized by the digit "1" and falsehood by the digit "0" . The connectives "and" , "or" , "if ... then" , and "is equivalent to" are denoted, respectively, by the symbols & (conjunction), ∨ (disjunction), → (implication), and ~ (equivalence); negation is represented by a bar □ above the symbol. In addition to individual propositions, examples of which are given above, one also uses variable propositions, i.e. variables whose values may be any individual statements given in advance. As the next step, the concept of a formula was inductively introduced. It formalizes the concept of a compound proposition. Let $A, B, C, \ldots$, denote individual propositions, and let $x, y, z, \ldots$, denote variable propositions. Each one of the above letters is known as a formula. Let the symbol * denote any one of the connectives listed, and let $\mathfrak{A}$ and $\mathfrak{B}$ denote formulas; then $(\mathfrak{A} * \mathfrak{B})$ and $\mathfrak{A}$ will be formulas (e.g. $((x\&y) \to \bar{z})$). The connectives and the particle "not" are looked upon as operations on quantities which assume the values 0 and 1, the results of these operations are also the digits 0 and 1. The conjunction $x\&y$ is equal to 1 if and only if both x and y equal 1; the disjunction $x \vee y$ equals 0 if and only if x and y are both equal to 0; the implication $x \to y$ is equal to 0 if and only if $x=1$ and $y=0$; the equivalence $x \sim y$ equals 1 if and only if the values of x and y are identical; and the negation $\bar{x}$ is equal to 1 if and only if $x=0$. If the values of the propositions in each formula are given, the value 0 or 1 can be assigned to the formula. This means that any formula can be considered as a way of stating or as a way of realizing a function of the algebra of logic, i.e. a function which is defined on samples of zeros and ones, and with as values also 0 or 1. Two formulas $\mathfrak{A}$ and $\mathfrak{B}$ are said to be equal $(\mathfrak{A} = \mathfrak{B})$ if they realize equal functions. The subject matter of the algebra of logic is the treatment of functions of the algebra of logic and the operations on these functions. In what follows below, the class of functions of the algebra of logic will be extended to the class of functions whose arguments, as well as the functions themselves, assume the values of elements of a finite fixed set E; the number of operations on the functions will also be extended. The algebra of logic is sometimes understood to mean this latter concept. However, from the aspect of practical applications, the most important case is that of sets E of cardinality two, and this case will therefore be discussed in greatest detail. The results presented here are also closely connected with a second approach to the study of propositions — the so-called propositional calculus.

In specifying the functions of the algebra of logic one often uses tables which contain all the combinations of the values of the variables and the values of the functions on these combinations. Thus, the synoptic table of the functions $\bar{x}$, $x\&y$, $x \vee y$, $x \to y$, and $x \sim y$ is given below:

x	y	$\bar{x}$	x&y	x∨y	x→y	x~y
0	0	1	0	0	1	1
0	1	1	0	1	1	0
1	0	0	0	1	0	0
1	1	0	1	1	1	1

Tables for arbitrary functions of the algebra of logic are constructed in a similar manner. This is the so-called tabular way of specifying functions of the algebra of logic. The tables themselves are sometimes called truth tables. The following equalities play an important part in the transformations of formulas into equivalent formulas:

$$x\&y = y\&x, \qquad x \vee y = y \vee x$$

(the law of commutativity);

$$(x\&y)\&z = x\&(y\&x), \qquad (x \vee y) \vee z = x \vee (y \vee x)$$

(the law of associativity);

$$x\&(x \vee y) = x, \qquad x \vee (x\&y) = x \text{ (the law of absorption);}$$

$$x \& (y \vee z) = (x\&y) \vee (x\&z),$$
$$x \vee (y\&z) = (x \vee y) \& (x \vee z)$$

(the laws of distributivity);

$$x \& \overline{x} = 0$$

(the law of contradiction);

$$x \vee \overline{x} = 1$$

(the law of the excluded middle);

$$x \to y = \overline{x} \vee y,$$
$$x \sim y = (x\&y) \vee (\overline{x} \& \overline{y}).$$

If these equalities are used, it is possible to obtain new equalities, even without the use of tables. These new equalities are obtained by the so-called identity transformations which, generally speaking, alter the expression, but not the function realized by the expression. For instance, the absorption laws yield the law of idempotency $x \vee x = x$. The above equalities frequently simplify the notation of formulas by eliminating some of the parentheses. Thus, the relations (1) and (2) make it possible to replace the formulas $((\ldots(\mathfrak{A}_1 \& \mathfrak{A}_2)\&\ldots)\&\mathfrak{A}_s)$ and $((\ldots(\mathfrak{B}_1 \vee \mathfrak{B}_2) \vee \ldots) \vee \mathfrak{B}_s)$ by the more compact notation $\mathfrak{A}_1 \& \mathfrak{A}_2 \& \ldots \& \mathfrak{A}_s$ and $\mathfrak{B}_1 \vee \mathfrak{B}_2 \vee \ldots \vee \mathfrak{B}_s$. The former relation is known as the conjunction of the factors $\mathfrak{A}_1, \ldots, \mathfrak{A}_s$, while the latter is called the disjunction of the terms $\mathfrak{B}_1, \ldots, \mathfrak{B}_s$. It is also seen from the equalities equations that, if the constants 0 and 1, the implication and the equivalence, are regarded as functions, they may be expressed by conjunctions, disjunctions and negations. Thus, any function of the algebra of logic may be realized by a formula written using only the symbols &, $\vee$ and □.

The set of all formulas involving variable propositions, some of the symbols &, $\vee$, $\to$, $\sim$, □, and the constants 0 and 1 is called a language over these symbols and constants. Equations show that, for any formula in the language over &, $\vee$, $\to$, $\sim$, □, 0, and 1, it is possible to find an equivalent formula in the language over &, $\vee$, □, 0, and 1; for example

$$(x \to y) \sim z = \left(\left(\overline{x} \vee y \right) \& z \right) \vee \left(\overline{\left(\overline{x} \vee y \right)} \& z \right)$$

In the latter language a special role is played by a class of formulas which may be written in the form $\mathfrak{A}_1 \vee \square \ldots \vee \mathfrak{A}_s$, 0 or 1, where $s \geq 1$ and where each $\mathfrak{A}_i$ is either a variable proposition, its negation or a conjunction of them, so that none of the $\mathfrak{A}_i$ contains equal factors or factors of the form x and $\overline{x}$ together, and all $\mathfrak{A}_i$ are pairwise unequal. Here the parentheses are omitted, since it is assumed that the conjunction operation is "stronger" than disjunction, i.e. when calculating with respect to the given values of the variables, the values of $\mathfrak{A}_1, \ldots, \mathfrak{A}_s$ are calculated first. These expressions are called disjunctive normal forms. Each formula u in a language over $\& \vee$, $\to$, $\sim$□, 0, 1 which realizes a non-zero function of the algebra of logic, may be converted with the aid of the equalities equations into an equivalent disjunctive normal form, containing all variable formulas of u and any number of other variables, and each $\mathfrak{A}_i$ in this disjunctive normal form containing the same variables. Such a disjunctive normal form is said to be a perfect disjunctive normal form of the formula u; the perfect disjunctive normal form of 0 is the formula 0 itself. The possibility of reduction to a perfect disjunctive normal form forms the basis of an algorithm determining the equality or inequality of two given formulas. The procedure of this algorithm is as follows: The given formulas u_1 and u_2 are reduced to perfect disjunctive normal forms which contain all the variables comprised in both u_1 and u_2 and the two expressions are compared; if they are identical, $u_1 = u_2$; if not, $u_1 \neq u_2$. An important role in the algebra of logic and its applications is played by contracted disjunctive normal forms, i.e. disjunctive normal forms which satisfy the following conditions: 1) it does not contain pairs of elements $\mathfrak{A}_i$ and $\mathfrak{A}_j$ such that any factor of $\mathfrak{A}_i$ is contained in $\mathfrak{A}_j$; 2) for any two elements $\mathfrak{A}_i$ and $\mathfrak{A}_j$ one of which contains some variable as factor, while the other contains the negation of this variable (under the condition that the pair of elements does not contain a second variable for which the same is true), there exists (in the same disjunctive normal form) an element $\mathfrak{A}_k$ which is equal to the conjunction of the other factors of these two elements. Any disjunctive normal form may be reduced with the aid of the equalities equations to an equal contracted disjunctive normal form. Thus, the contracted disjunctive normal form for the formula $\left(\left(x \sim (y \rightarrow z)\right) \rightarrow (x \& z)\right)$ is $\bar{x} \& \bar{y} \vee z \vee x \& y$. The formulas u_1 and u_2 are equal if and only if their contracted disjunctive normal forms are identical. In addition to the disjunctive normal form, use is also made of the conjunctive normal form, i.e. of expressions which are obtained from a disjunctive normal form by substituting the symbol $\vee$ for $\&$, $\&$ for $\vee$, and 0 for 1. For instance, the disjunctive normal form $x \& y \vee \bar{x} \& z$ yields the conjunctive normal form $(x \vee y) \& (\bar{x} \vee z)$. An operation (or function) f is said to be the dual of an operation ψ if the table defining f is obtained from the table which defines ψ by substituting 1 for 0 and 0 for 1 throughout (including interchange in the values of the functions).

Thus, the conjunction and the disjunction are mutually dual, the negation is dual to itself, the constants 0 and 1 are mutually dual, etc. A transformation of a formula which consists in replacing the symbols of all operations and expressions by the symbols of their dual operations, and in exchanging 0 for 1 and 1 for 0, is called a duality transformation. If the equality $\mathfrak{A} = \mathfrak{B}$ is true, and if $\mathfrak{A}^*$ is the dual of $\mathfrak{A}$ while $\mathfrak{B}^*$ is the dual of $\mathfrak{B}$, then it is also true that $\mathfrak{A}^* = \mathfrak{B}^*$, and this equality is called the dual of the above equality; this is the so-called duality principle. The pairs of the laws are examples of dual equalities; equality is dual to equality and each conjunctive normal form is the dual of some disjunctive normal form. A perfect conjunctive normal form and a contracted conjunctive normal form are defined as conjunctive normal forms such that their dual expressions are a perfect disjunctive normal form and a contracted disjunctive normal form, respectively. Perfect and contracted disjunctive and conjunctive normal forms are used for surveying all hypotheses and all consequences of a given formula. A hypothesis of a formula $\mathfrak{A}$ is understood to be a formula $\mathfrak{B}$ such that $(\mathfrak{B} \rightarrow \mathfrak{A}) = 1$; a consequence of a formula $\mathfrak{A}$ is a formula $\mathfrak{B}$ such that $(\mathfrak{A} \rightarrow \mathfrak{B}) = 1$. A hypothesis of a formula $\mathfrak{A}$ is said to be simple if it is a conjunction of variable formulas or their negations and if, after discarding any one of its factors, it is no longer a hypothesis of the formula $\mathfrak{A}$. Similarly, a consequence of $\mathfrak{A}$ is called simple if it is a disjunction of variable formulas or their negations and if, after one of its elements is discarded, it is no longer a consequence of $\mathfrak{A}$.

The survey of hypotheses and consequences is based on the indication of an algorithm which transforms a given formula into all of its simple hypotheses and consequences and, with the aid of the laws, into all remaining hypotheses and consequences. The algorithm is based on the following facts. If $\mathfrak{A} = \mathfrak{B}$, then $\mathfrak{A}$ and $\mathfrak{B}$ have identical hypotheses and identical consequences. An element of a disjunctive normal form is a hypothesis of that disjunctive normal form, while a factor of a conjunctive normal form is a consequence of that conjunctive normal form. If $\mathfrak{A}$ is a hypothesis of a proposition $\mathfrak{B}$, then $\mathfrak{A}\&\mathfrak{C}$ is also a hypothesis for $\mathfrak{B}$; if $\mathfrak{A}$ is a consequence of $\mathfrak{B}$, then $\mathfrak{A} \vee \mathfrak{C}$ is also a consequence of $\mathfrak{B}$. If $\mathfrak{A}$ and $\mathfrak{C}$ are hypotheses of a proposition $\mathfrak{B}$, then $\mathfrak{A} \vee \mathfrak{C}$ is also a hypothesis for $\mathfrak{B}$; if $\mathfrak{A}$ and $\mathfrak{C}$ are consequences of $\mathfrak{B}$, then $\mathfrak{A}\&\mathfrak{C}$ is also a consequence of $\mathfrak{B}$. A perfect disjunctive normal form has no hypotheses (not containing letters which are not contained in that disjunctive normal form) other than the disjunctions of some of its elements or of disjunctive normal forms equal to them.

A perfect conjunctive normal form has no consequences other than the conjunctions of some of its factors or of propositions equal to them. A contracted disjunctive normal form is the disjunction of all its simple hypotheses; a contracted conjunctive normal form is the conjunction of all its simple consequences. A contracted disjunctive normal form has important applications. The first problem which should be mentioned is the minimization of functions of the algebra of logic, which is a part of the problem of synthesis of control (switching) systems. The minimization of functions of the algebra of logic consists in constructing a disjunctive normal form for the given function of the algebra of logic which realizes this function and has the smallest sum of the number of factors in its elements, i.e. has minimal complexity. Such disjunctive normal forms are called minimal. Each minimal disjunctive normal form for a given function of the algebra of logic which is not a constant is obtained from the contracted disjunctive normal form of this function by discarding a number of elements. For certain functions the contracted disjunctive normal form may coincide with the minimal disjunctive normal form. Such a case occurs, for example, with monotone functions, i.e. functions realizable by formulas over $\&, \vee, 0, 1$.

In the language over $\&, \vee, \rightarrow, \sim, 0, 1, +$, when the sign $+$ is interpreted as addition modulo 2, the following relations are valid:

$$x \vee y = \left(\left(x\&y \right) + x \right) + y,$$

$$x \rightarrow y = \overline{x}\&y, \qquad x \sim y = \left(x+y \right) + 1,$$

$$x+y = \left(x\&y \right) \vee \left(\overline{x}\&\overline{y} \right), \qquad 1 = x \vee \overline{x}.$$

These formulas make it possible to translate formulas from the language over $\&, \vee, \rightarrow, \overline{\Box}, 0, 1$ into equivalent formulas in the language over $\&, +, 1$, and vice versa. The identity transformations in the latter language are realized with the aid of equalities established for the conjunction and the following additional equalities:

$$x+y = y+x,$$

$$\left(x+y \right) + z = x + \left(y+z \right),$$

$$x\& \left(y+z \right) = x\&y + x\&z,$$

$$x\&y = x, \qquad x + \left(y+y \right) = x, \qquad x\&1 = x$$

Here it is considered, as before, that conjunction is a stronger connective than the sign $+$. These equalities are sufficient to deduce any valid equality in the language over $\&, +, 1$, with the aid of identity transformations, just as in the case of the language over $\&, \vee, \rightarrow, \sim, \square, +1$. A proposition in this language is called a reduced polynomial if it is of the form $\mathfrak{A}_1 + \ldots + \mathfrak{A}_s$, when $\mathfrak{A}_i$ is equal to 1 or is a variable or a conjunction of various variables without negations, $\mathfrak{A}_i \neq \mathfrak{A}_j$ if $i \neq j$, $s \geq 1$, or else is equal to $1+1$. Thus, the expression $x\&y\&z+x\&y+1$ is a reduced polynomial. Any formula of the algebra of logic may be converted to a reduced polynomial by identity transformations. The equality $\mathfrak{A} = \mathfrak{B}$ is valid if and only if the reduced polynomial for $\mathfrak{A}$ is identical with the reduced polynomial for $\mathfrak{B}$. In addition to the languages mentioned above there are also other languages equivalent to them. Two languages are called equivalent if each formula in one language can be converted, by means of certain conversion rules, into another equivalent formula in the second language, and vice versa. It is sufficient to base such a language on an arbitrary system of operations (and constants) such that the operations (and constants) of the system may be used to represent any function of the algebra of logic. Such systems are said to be functionally complete. Examples of complete systems are $\{\overline{x \vee y}\}, \{x \vee y, \overline{x}\}, \{x+y, 1, x\&y\}$ etc. There exists an algorithm which can be used to establish the completeness or incompleteness of an arbitrary finite system of functions of the algebra of logic. It is based on the following fact. A system of functions of the algebra of logic is complete if and only if it contains functions $f_1(x, y,...,v)$ and $f_2(x, y,...,v)$ such that $f_1(0,0,...,0 = 1)$ and $f_2(1,1,...,1) = 0$, as well as functions f_3, f_4 and f_5, where $f_3 \neq f_3^*$, f_4 is not monotone, and the reduced polynomial of f_5 contains an element $\mathfrak{A}$ with more than one factor. There are also other languages based on systems of operations which are not functionally complete, and the number of such languages is infinite. They contain infinitely many pairwise incomparable languages (in the sense that there is no way of translating from one language to another by means of identity transformations). However, for any language based on some operations of the algebra of logic there exists a finite system of equations in this language such that any equality can be deduced with the aid of identity transformations of the equations of this system. Such a system called a deductively complete system of equalities in this language. Thus, the equalities constitute a complete system of equalities of the language over $\& \vee, \square, 0, 1$.

In considering one of the languages discussed above in conjunction with some complete system of equalities in this language, a tabular statement of the basic operations in the language and the requirement that propositions be the values of its variables are sometimes discarded. Instead, various interpretations of the language are permitted. These interpretations involve some set of objects (which are used as values of the variables) and some system of operations over the objects of this set, which satisfy the equalities from a complete system of equalities of the language. Thus, such a procedure results in the transformation of the language over $\& \vee, \square, 0, 1$ into the language of a Boolean algebra; the language over $\&, +, 1$ is transformed into the language of a Boolean ring (with a unit element); the language over $\&, \vee, \overline{\square}$ is transformed into the language of a distributive lattice, etc.

Historically, the development of the algebra of logic has been stimulated mainly by the problems to which it has been applied. An important application is the theory of electric circuits. In certain cases circuits cannot be described in terms of the ordinary two-valued algebra of logic, and a many-valued logic must be considered.

Boolean Algebra

Boolean logic is a form of algebra in which all values are reduced to either TRUE or FALSE. Boolean logic is especially important for computer science because it fits nicely with the binary numbering system, in which each bit has a value of either 1 or 0. Another way of looking at it is that each bit has a value of either TRUE or FALSE.

There are several more Boolean operators besides just AND. The most common ones are AND, OR, NOT, NAND, NOR, XOR. Let's go over them really quickly. We'll use the conventional letters P and Q.

AND means that if P and Q are both true, the result is true. Otherwise, the result is false.

OR means that if either P or Q are true, the result is true. Otherwise, the result is false.

NOT means to flip the current state. True becomes false and false becomes true.

NAND is just like AND but you flip each possible combination.

NOR is just like OR but you flip each possible combination.

XOR means if either P or Q is true - but not both - the output is true. Otherwise, the result is false.

Maybe we should show you some truth tables? They make it super simple. For any combination, just line up the present states of P and Q and read across.

P	Q	P AND Q	P	Q	P OR Q	P	Q	P XOR Q
F	F	F	F	F	F	F	F	F
F	T	F	F	T	T	F	T	T
T	F	F	T	F	T	T	F	T
T	T	T	T	T	T	T	T	F

P	Q	P NAND Q	P	Q	P NOR Q	P	Q	NOT(P)	NOT(Q)
F	F	T	F	F	T	F	F	T	T
F	T	T	F	T	F	F	T	T	F
T	F	T	T	F	F	T	F	F	T
T	T	F	T	T	F	T	T	F	F

Some commonly used truth tables

A set of rules or Laws of Boolean Algebra expressions have been invented to help reduce the number of logic gates needed to perform a particular logic operation resulting in a list of functions or theorems known commonly as the Laws of Boolean Algebra.

Boolean Algebra is the mathematics we use to analyse digital gates and circuits. We can use these "Laws of Boolean" to both reduce and simplify a complex Boolean expression in an attempt to reduce the number of logic gates required. *Boolean Algebra* is therefore a system of mathematics based on logic that has its own set of rules or laws which are used to define and reduce Boolean expressions.

The variables used in Boolean Algebra only have one of two possible values, a logic "0" and a logic "1" but an expression can have an infinite number of variables all labelled individually to represent inputs to the expression, For example, variables A, B, C etc, giving us a logical expression of A + B = C, but each variable can ONLY be a 0 or a 1.

Examples of these individual laws of Boolean, rules and theorems for Boolean Algebra are given in the following table.

Truth Tables for the Laws of Boolean

Boolean Expression	Description	Equivalent Switching Circuit	Boolean Algebra Law or Rule
A + 1 = 1	A in parallel with closed = "CLOSED"		Annulment
A + 0 = A	A in parallel with open = "A"		Identity
A . 1 = A	A in series with closed = "A"		Identity
A . 0 = 0	A in series with open = "OPEN"		Annulment
A + A = A	A in parallel with A = "A"		Idempotent
A . A = A	A in series with A = "A"		Idempotent
NOT A = A	NOT NOT A (double negative) = "A"		Double Negation
A + $\overline{A}$ = 1	A in parallel with NOT A = "CLOSED"		Complement
A . $\overline{A}$ = 0	A in series with NOT A = "OPEN"		Complement
A+B = B+A	A in parallel with B = B in parallel with A		Commutative
A.B = B.A	A in series with B = B in series with A		Commutative
$\overline{A+B}$ = $\overline{A}$.$\overline{B}$	invert and replace OR with AND		de Morgan's Theorem
$\overline{A.B}$ = $\overline{A}$+$\overline{B}$	invert and replace AND with OR		de Morgan's Theorem

The basic Laws of Boolean Algebra that relate to the Commutative Law allowing a change in position for addition and multiplication, the Associative Law allowing the removal of brackets for addition and multiplication, as well as the Distributive Law allowing the factoring of an expression, are the same as in ordinary algebra.

Each of the *Boolean Laws* above are given with just a single or two variables, but the number of variables defined by a single law is not limited to this as there can be an infinite number of variables as inputs too the expression. These Boolean laws detailed above can be used to prove any given Boolean expression as well as for simplifying complicated digital circuits.

A brief description of the various Laws of Boolean are given below with A representing a variable input.

Description of the Laws of Boolean Algebra

- Annulment Law – A term AND´ed with a "0" equals 0 or OR´ed with a "1" will equal 1

  $A . 0 = 0$ A variable AND'ed with 0 is always equal to 0

  $A + 1 = 1$ A variable OR'ed with 1 is always equal to 1

- Identity Law – A term OR´ed with a "0" or AND´ed with a "1" will always equal that term

  $A + 0 = A$ A variable OR'ed with 0 is always equal to the variable

  $A . 1 = A$ A variable AND'ed with 1 is always equal to the variable

- Idempotent Law – An input that is AND´ed or OR´ed with itself is equal to that input

  $A + A = A$ A variable OR'ed with itself is always equal to the variable

  $A . A = A$ A variable AND'ed with itself is always equal to the variable

- Complement Law – A term AND´ed with its complement equals "0" and a term OR´ed with its complement equals "1"

  $A . \overline{A} = 0$ A variable AND'ed with its complement is always equal to 0

  $A + \overline{A} = 1$ A variable OR'ed with its complement is always equal to 1

- Commutative Law – The order of application of two separate terms is not important

  $A . B = B . A$ The order in which two variables are AND'ed makes no difference

  $A + B = B + A$ The order in which two variables are OR'ed makes no difference

- Double Negation Law – A term that is inverted twice is equal to the original term

  $\overline{\overline{A}} = A$ A double complement of a variable is always equal to the variable

- de Morgan´s Theorem – There are two "de Morgan´s" rules or theorems,

  (1) Two separate terms NOR´ed together is the same as the two terms inverted (Complement) and AND´ed for example: $\overline{A+B} = \overline{A} . \overline{B}$

(2) Two separate terms NAND´ed together is the same as the two terms inverted (Complement) and OR´ed for example: $\overline{A.B} = \overline{A} + \overline{B}$.

Other algebraic Laws of Boolean not detailed above include:

- Distributive Law – This law permits the multiplying or factoring out of an expression.

  $A(B + C) = A.B + A.C$ (OR Distributive Law)

  $A + (B.C) = (A + B).(A + C)$ (AND Distributive Law)

- Absorptive Law – This law enables a reduction in a complicated expression to a simpler one by absorbing like terms.

  $A + (A.B) = A$ (OR Absorption Law)

  $A(A + B) = A$ (AND Absorption Law)

- Associative Law – This law allows the removal of brackets from an expression and regrouping of the variables.

  $A + (B + C) = (A + B) + C = A + B + C$ (OR Associate Law)

  $A(B.C) = (A.B)C = A . B . C$ (AND Associate Law)

Boolean Algebra Functions

Using the information above, simple 2-input AND, OR and NOT Gates can be represented by 16 possible functions as shown in the following table.

Function	Description	Expression
1.	NULL	0
2.	IDENTITY	1
3.	Input A	A
4.	Input B	B
5.	NOT A	$\overline{A}$
6.	NOT B	$\overline{B}$
7.	A AND B (AND)	A.B
8.	A AND NOT B	$A.\overline{B}$
9.	NOT A AND B	$\overline{A}.B$
10.	NOT AND (NAND)	$\overline{A.B}$
11.	A OR B (OR)	A + B
12.	A OR NOT B	$A + \overline{B}$
13.	NOT A OR B	$\overline{A} + B$
14.	NOT OR (NOR)	$\overline{A+B}$
15.	Exclusive-OR	$A.\overline{B} + \overline{A}.B$
16.	Exclusive-NOR	$A.B + \overline{A.B}$

Laws of Boolean Algebra Example No1

Using the above laws, simplify the following expression: $(A + B)(A + C)$

$Q =$	$(A + B).(A + C)$	
	$A.A + A.C + A.B + B.C$	– Distributive law
	$A + A.C + A.B + B.C$	– Idempotent AND law $(A.A = A)$
	$A(1 + C) + A.B + B.C$	– Distributive law
	$A.1 + A.B + B.C$	– Identity OR law $(1 + C = 1)$
	$A(1 + B) + B.C$	– Distributive law
	$A.1 + B.C$	– Identity OR law $(1 + B = 1)$
$Q =$	$A + (B.C)$	– Identity AND law $(A.1 = A)$

Then the expression: $(A + B)(A + C)$ can be simplified to $A + (B.C)$ as in the Distributive law.

Bunched Logic

A bunched logic is a logic in which the formulas in the context are not just a list or set but have some additional, usually tree-like, structure.

This can be indicated syntactically by the use of two or more punctuation symbols, such as comma and semicolon, along with parentheses for grouping. Thus for instance a sequent with bunches might be written like

$$A,B,(C;(D,E);F),(G;H) \vdash K$$

The contexts put together with both commas and semicolons are called bunches. The general phrase bunched logic is not entirely standard, although the word "bunches" has been used with more than one logic of this form.

Each type of punctuation also comes with a nullary version. The punctuation symbols like comma and semicolon are sometimes called structural connectives, since they are part of the judgmental structure (like structural rules) but are closely related to the logical connectives such as conjunction and disjunction.

Usually the reason for using a bunched logic is that the different structural connectives obey different structural rules. For instance, the semicolon might allow the contraction rule and/or the weakening rule, while the comma does not.

Examples

Bunched Implications

In bunched implications logic (BI), the semicolon allows contraction and weakening while the comma does not. This allows defining both an additive conjunction and a multiplicative conjunction, internalizing the semicolon and the comma respectively, such that both distribute over the additive disjunction and moreover both come with a corresponding implication: the additive implication? and the multiplicative implication.

Relevance Logic

Some forms of relevance logic can be presented with a bunched sequent calculus that is similar to BI, but in which the comma also has contraction (though not weakening) and there is no additive implication.

Classical Bunched Implication

A logic of classical bunched implication is like BI but with arbitrary bunches on the right as well as on the left. On the right, the semicolon represents the additive disjunction and the comma represents a multiplicative disjunction, and there are both an additive and a multiplicative negation that move formulas back and forth. Both negations are "classical" with respect to their corresponding connectives, e.g. we have $\sim\sim A \multimap A$ (where $\sim$ is the multiplicative negation and $\multimap$ the multiplicative implication) and also $\neg\neg A \to A$ (where $\neg$ is the additive negation and $\to$ the additive implication).

Categorical Semantics

Bunched logics naturally have semantics in categories with more than one monoidal structure, so that a bunch such as $(A,(B;C),((D,E);F))$ can be interpreted as $A \otimes (B \boxtimes C) \otimes ((D \otimes E) \boxtimes F)$. Frequently (e.g. if one kind of bunch admits contraction and weakening) one of the two monoidal structures is a cartesian one. A typical and motivating example of a model for BI is the category of presheaves $[C^{op}, Set]$ over a monoidal category C, which comes equipped both with the ordinary ccc structure on presheaves as well as the closed monoidal structure given by Day convolution.

Applications

Interference Control

In perhaps the first use of substructural type theory to control resources, John C. Reynolds showed how to use an affine type theory to control aliasing and other forms of interference in Algol-like programming languages. O'Hearn used bunched type theory to extend Reynolds system by allowing interference and non-interference to be more flexibly mixed. This resolved open problems concerning recursion and jumps in Reynolds's system.

Separation Logic

Separation logic is an extension of Hoare logic which facilitates reasoning about mutable data structures that use pointers. Following Hoare logic the formulae of separation logic are of the form $\{Pre\}program\{Post\}$, but the preconditions and postconditions are formulae interpreted in a model of bunched logic. The original version of the logic was based on models as follows:

- $Heaps = L \rightharpoonup_f V$ (finite partial functions from locations to values)

- $h_0 \bullet h_1$ = union of heaps with disjoint domains, undefined when domains overlap.

It is the undefinedness of the composition on overlapping heaps that models the separation idea. This is a model of the boolean variant of bunched logic.

Separation logic was used originally to prove sequential programs, but then was extended to concurrency using a proof rule

$$\frac{\{P_1\}C_1\{Q_1\} \quad \{P_2\}C_2\{Q_2\}}{\{P_1 * P_1\}C_1 \| C_2\{Q_1 * Q_2\}}$$

that divides the storage accessed by parallel threads.

Later, the greater generality of the resource semantics was utilized: an abstract version of separation logic works for Hoare triples where the preconditions and postconditions are formulae interpreted over an arbitrary partial commutative monoid instead of a particular heap model. By suitable choice of commutative monoid, it was surprisingly found that the proofs rules of abstract versions of concurrent separation logic could be used to reason about interfering concurrent processes, for example by encoding rely-guarantee and trace-based reasoning.

Separation logic is the basis of a number of tools for automatic and semi-automatic reasoning about programs, and is used in the Infer program analyzer currently deployed at Facebook.

Resources and Processes

Bunched logic has been used in connection with the (synchronous) resource-process calculus SCRP in order to give a (modal) logic which characterizes, in the sense of Hennessey-Milner, the compositional structure of concurrent systems.

SCRP is notable for interpreting $A * B$ in terms of *both* parallel composition of systems and composition of their associated resources. The semantic clause of SCRP's process logic that corresponds to separation logic's rule for concurrency asserts that a formula $A * B$ is true in resource-process state R, E just in case there are decompositions of the resource $R = S \bullet T$ and process $E \sim F \times G$, where $\sim$ denotes bisimulation, such that A is true in the resource-process state S, F and B is true in the resource-process state T, G; that is $R, E \vDash A$ iff $S, F \vDash A$ and $T, G \vDash B$.

The system SCRP is based directly on bunched logic's resource semantics; that is, on ordered monoids of resource elements. While direct and intuitively appealing, this choice leads to a specific technical problem: the Hennessy-Milner completeness theorem holds only for fragments of the modal logic that exclude the multiplicative implication and multiplicative modalities. This problem is solved by basing resource-process calculus on a resource semantics in which resource elements are combined using two combinators, one corresponding to concurrent composition and one corresponding to choice.

Spatial Logics

Cardelli, Caires, Gordon and others have investigated a series of logics of process calculi, where a conjunction is interpreted in terms of parallel composition. [References, to add] Unlike the work of Pym et al. in SCRP, they do not distinguish between parallel composition of systems and composition of resources accessed by the systems.

Their logics are based on instances of the resource semantics which give rise to models of the boolean variant of bunched logic. Although these logics give rise to instances of boolean bunched logic, they appear to have been arrived at independently, and in any case have significant additional structure in the way of modalities and binders. Related logics have been proposed as well for modelling XML data.

Classical Logic

In classical logic, a simple proposition P is a linguistic, or declarative, statement contained within a universe of elements, X, that can be identified as being a collection of elements in X that are strictly true or strictly false.

The veracity (truth) of an element in the proposition P can be assigned a binary truth value, called T (P),

For binary (Boolean) classical logic, T (P) is assigned a value of 1 (truth) or 0 (false).

If U is the universe of all propositions, then T is a mapping of the elements, u, in these propositions (sets) to the binary quantities (0, 1), or

$$T : u \in U \longrightarrow (0, 1)$$

Example: Let P be the proposition "The structural beam is an 18WF45" and let Q be the proposition "The structural beam is made of steel." Let X be the universe of structural members comprised of girders, beams, and columns; x is an element (beam), A is the set of all wide-flange (WF) beams, and B is the set of all steel beams. Hence,

> P : x is in A

> Q : x is in B

Let P and Q be two simple propositions on the same universe of discourse that can be combined using the following five logical connectives

> Disjunction (∨)

> Conjunction (∧)

> Negation (−)

> Implication (→)

> Equivalence (↔)

define sets A and B from universe, where these sets might represent linguistic ideas or thoughts.

A propositional calculus (sometimes called the algebra of propositions) will exist for the case where proposition P measures the truth of the statement that an element, x, from the universe X is contained in set A and the truth of the statement Q that this element, x, is contained in set B, or more conventionally,

P : truth that $x \in A$

Q : truth that $x \in B$

where truth is measured in terms of the truth value, i.e.,

İf $x \in A$, $T(P) = 1$; otherwise, $T(P) = 0$

İf $x \in B$, $T(Q) = 1$; otherwise, $T(Q) = 0$

or, using the characteristic function to represent truth (1) and falsity (0), the following notation results:

$$xA(x) = \begin{cases} 1, & x \notin A \\ 0, & x \notin A \end{cases}$$

The five logical connectives already defined can be used to create compound propositions, where a compound proposition is defined as a logical proposition formed by logically connecting two or more simple propositions.

Disjunction

$P \vee Q : x \in A$ or $x \in B$

Hence, $T(P \vee Q) = \max(T(P), T(Q))$

Conjunction

$P \wedge Q : x \in A$ and $x \in B$

Hence, $T(P \wedge Q) = \min(T(P), T(Q))$

Negation

If $T(P) = 1$, then $T(\overline{P}) = 0$; if $T(P) = 0$, then $T(\overline{P}) = 1$.

Implication

$(P \rightarrow Q) : x \notin A$ or $x \in B$

Hence, $T(P \rightarrow Q) = T(\overline{P} \cup Q)$

Equivalence

$$(P \leftrightarrow Q) : T(P \leftrightarrow Q) = \begin{cases} 1, & \text{for } T(P) = T(Q) \\ 0, & \text{for } T(P) \neq T(Q) \end{cases}$$

P	Q	$\overline{P}$	P∨Q	P∧Q	P→Q	P↔Q
T(1)	T(1)	F(0)	T(1)	T(1)	T(1)	T(1)
T(1)	F(0)	F(0)	T(1)	F(0)	F(0)	F(0)
F(0)	T(1)	T(1)	T(1)	F(0)	T(1)	F(0)
F(0)	F(0)	T(1)	F(0)	F(0)	T(1)	T(1)

The implication P →Q can be represented in set-theoretic terms by the relation R,

Suppose the implication operation involves two different universes of discourse; P is a proposition described by set A, which is defined on universe X, and Q is a proposition described by set B, which is defined on universe Y.

$$R=\left(A\times B\right)\cup\left(\overline{A}\times Y\right)\equiv \text{IF A, THEN B}$$
$$\text{IF } x\in A \text{ where } x\in X \text{ and A } \subset X$$
$$\text{THEN } y\in B \text{ where } y\in Y \text{ and } B\subset Y$$

This implication is also equivalent to the linguistic rule form,

If A, then B.

$$P\to Q: \text{IF } x\in A, \text{THEN } y\in B, \text{ or } P\to Q\equiv \overline{A}\cup B$$

Another compound proposition in linguistic rule form is the expression

If A, Then B, Else C

The set-theoretic equivalent of this compound proposition is given by

$$\text{IF A, THEN B, ELSE C } \equiv \left(A\times B\right)\cup\left(\overline{A}\times C\right)=R= \text{relation on X}\times Y$$

Tautologies

In classical logic it is useful to consider compound propositions that are always true, irrespective of the truth values of the individual simple propositions.

Classical logical compound propositions with this property are called *tautologies*.

Tautologies are useful for deductive reasoning, for proving theorems, and for making deductive inferences.

Some common tautologies follow:

$$\overline{B}\cup B \leftrightarrow X$$
$$A\cup X; \ \overline{A}\cup X \leftrightarrow X$$

$$\left(A \wedge (A \rightarrow B)\right) \rightarrow B \quad (modus\ ponens)$$

$$\left(\overline{B} \wedge (A \rightarrow B)\right) \rightarrow \overline{A} \quad (modus\ tollens)$$

Truth table (*modus ponens*)

A	B	A→B	(A∧(A→B))	(A∧(A→B))→B	
0	0	1	0	1	
0	1	1	0	1	Tautology
1	0	0	0	1	
1	1	1	1	1	

Truth table (*modus tollens*)

A	B	$\overline{A}$	$\overline{B}$	A→B	$(\overline{B} \wedge (A \rightarrow B))$	$(\overline{B} \wedge (A \rightarrow B)) \rightarrow \overline{A}$	
0	0	1	1	1	1	1	
0	1	1	0	1	0	1	Tautology
1	0	0	1	0	0	1	
1	1	0	0	1	0	1	

Deductive Inferences

The modus ponens deduction is used as a tool for making inferences in rule-based systems. A typical if–then rule is used to determine whether an antecedent (cause or action) infers a consequent (effect or reaction).

Suppose we have a rule of the form IF A, THEN B, where A is a set defined on universe X and B is a set defined on universe Y. As discussed before, this rule can be translated into a relation between sets A and B;

$$R = (A \times B) \cup \left(\overline{A} \times Y\right)$$

Suppose a new antecedent, say A', is known. Can we use modus ponens deduction to infer a new consequent, say B', resulting from the new antecedent? That is, can we deduce, in rule form, IF A', THEN B'?

Yes, through the use of the composition operation. Since "A implies B" is defined on the Cartesian space X × Y, B can be found through the following set-theoretic formulation,

$$B' = A' \circ R = A' \circ \left((A \times B) \cup \left(\overline{A} \times Y\right)\right)$$

The rule IF A, THEN B (proposition P is defined on set A in universe X, and proposition Q is defined on set B in universe), i.e., (P→Q) = R = (A X B) ∪ ($\overline{A}$ X Y), is then defined in function-theotic terms as

$$\chi R\left(x,y\right)=\max\left[\left(\chi A(x)\wedge\chi B(y)\right),\left(\left(1-\chi A(x)\right)\wedge 1\right)\right]$$

Where $\chi\,(\;)$ is the characteristic function as defined before.

The compound rule IF A, THEN B, ELSE C can also be defined in terms of a matrix relation as

$$R=\left(A\times B\right)\cup\left(\overline{A}\times C\right)\Rightarrow\left(P\to Q\right)\wedge\left(\overline{P}\to S\right)$$

where the membership function is determined as

$$\chi_{R}(x,y)=\max\left[\left(\chi A(x)\wedge\chi B(y)\right),\left(\left(1-\chi A(x)\right)\wedge\chi_{C}(y)\right)\right]$$

Example

Suppose we have two universes of discourse for a heat exchanger problem described by the following collection of elements,

   X = {1, 2, 3, 4} and

   Y = {1, 2, 3, 4, 5, 6}.

Suppose X is a universe of normalized temperatures and Y is a universe of normalized pressures.

Define crisp set A on universe X and crisp set B on universe Y as follows:

   A = {2, 3} and

   B = {3, 4}.

The deductive inference IF A, THEN B (i.e., IF temperature is A, THEN pressure is B) will yield a matrix describing the membership values of the relation R, i.e., $\chi R(x, y)$

That is, the matrix R represents the rule IF A, THEN B as a matrix of characteristic (crisp membership) values.

$$B'=A'\circ R=A'\circ\left(\left(A\times B\right)\cup\left(\overline{A}\times Y\right)\right).$$

Non-classical Logic

The purpose of this entry is to survey those modern logics that are often called "non-classical," classical logic being the theory of validity concerning truth functions and first-order quantifiers likely to be found in introductory textbooks of formal logic at the end of the twentieth century.

For the sake of uniformity I will give a model-theoretic account of the logics. All of the logics also have proof-theoretic characterizations, and in some cases (such as linear logic) these characterizations are somewhat more natural. I will not discuss combinatory logic, which is not so much a non-classical logic as it is a way of expressing inferences that may be deployed for both classical and non-classical logics. I will use A, B, ... for arbitrary sentences; $\land$, $\lor$, $\lnot$, and $\to$, for the standard conjunction, disjunction, negation, and conditional operators for whichever logic is at issue. "Iff" means "if and only if."

Extensions versus Rivals

An important distinction is that between those non-classical logics that take classical logic to be alright as far as it goes, but to need extension by the addition of new connectives, and those which take classical logic to be incorrect, even for the connectives it employs. Call the former *extensions* of classical logic, and the latter *rivals*. Thus modal logics, as now usually conceived, are extensions of classical logic. They agree with classical logic on the extensional connectives (and quantifiers if these are present) but augment them with modal operators. By contrast, intuitionist and relevant logics are more plausibly thought of as rivals. Thus $A \lor \lnot A$ is valid in classical logic but not intuitionist logic, and $A \to (B \to A)$ is valid in classical logic but not relevant logic.

An important distinction is that between those non-classical logics that take classical logic to be alright as far as it goes, but to need extension by the addition of new connectives, and those which take classical logic to be incorrect, even for the connectives it employs. Call the former *extensions* of classical logic, and the latter *rivals*. Thus modal logics, as now usually conceived, are extensions of classical logic. They agree with classical logic on the extensional connectives (and quantifiers if these are present) but augment them with modal operators. By contrast, intuitionist and relevant logics are more plausibly thought of as rivals. Thus $A \lor \lnot A$ is valid in classical logic but not intuitionist logic, and $A \to (B \to A)$ is valid in classical logic but not relevant logic.

The distinction must be handled with care however. Modern modal logics can be formulated, not with the modal operators, but with the strict conditional, $\dashv$ (from which modal operators can be defined), as primitive; and $A \dashv (B \dashv A)$ is not valid. From this perspective modal logic is a rival to classical logic (which is the way it was originally intended). Similarly it is (arguably) possible to add a negation operator, $, to relevant logics which behaves as does classical negation. Classical logic is, then, just a part of this logic, identifying the classical $\lnot A$ and $A \to B$ with the relevant A and $A \lor B$, respectively. From this perspective, in a relevant logic, $\to$ and $\lnot$ are operators additional to the classical ones, and relevant logic is an extension of classical logic.

What these examples show is that whether or not something is an extension or a rival of classical logic is not a purely formal matter but a matter of how the logic is taken to be applied to informal reasoning. If, in a modal logic, one reads $A \dashv B$ as "if A then B " then the logic is a rival of classical logic. If one reads $A \to B$ as "if A then B " and $A \dashv B$ as "necessarily, if A then B," it is an extension. If, in a relevant logic, one reads $A \to B$ as "if A then B," and $\lnot A$ as "it is not the case that A," the logic is a rival to classical logic; if one reads $A \lor B$ as "if A then B " and A as "it is not the case A," it is an extension. (The examples also raise substantial philosophical issues. Thus both a relevant logician and an intuitionist are liable to deny that $ is a connective with any determinate meaning.)

Many-Valued Logics

A central feature of classical logic is its bivalence. Every sentence is exclusively either true (1) or false (0). In *many-valued logics*, normally thought of as rivals to classical logic, there are more than two semantic values. Truth-functionality is, however, maintained; thus the value of a compound formula is determined by the values of its components. Some of the semantic values are *designated*, and a valid inference is one in which, whenever the premises are *designated*, so is the conclusion.

A simple example of a many-valued logic is that in which there are three truth values, 1, *i*, 0; and the truth functions for the standard connectives may be depicted as follows:

The only designated value is 1 (which is what the asterisk indicates). This is the Łukaziewicz 3-valued logic, $Ł_3$. If the middle value of the table for → is changed from 1 to *i* we get the Kleene 3-valued logic K_3. The standard interpretation for *i* in this logic is *neither true nor false*. If in addition *i* is added as a designated value, we get the paraconsistent logic *LP*. The standard interpretation for *i* in this is *both true and false*.

$Ł_3$ can be generalized to a logic, $Ł_n$, with *n* values, for any finite *n*, and even to one with infinitely many values. Thus the continuum-valued Łukasiewicz logic, $Ł_\aleph$, has as semantic values all real numbers between 0 and 1 (inclusive). Normally only 1 is designated. If we write the value of *A* as v(*A*), v(*A* ∨*B*) and v(*A* ∧*B*) are the maximum and minimum of v(*A*) and v(*B*), respectively; v(¬*A*)=1-v(*A*); v(*A* →*B*)=1 if v(*A*)≤v(*B*) and v(*A* →*B*)=1-(v(*A*)-v(*B*)) otherwise. Standardly the semantic values are thought of as degrees of truth (so that 1 is *completely true*). Interpreted in this way $Ł_\aleph$ is one of a family of many-valued logics called *fuzzy logics*.

Intensional Logics

World semantics have turned out to be one of the most versatile techniques in contemporary logic. Generally speaking, logics that have world-semantics are called *intensional logics* (and are normally thought of as extensions of classical logic). There are many of these in addition to standard modal logics.

□ may be interpreted as "it is known that", in which context it is usually written as *K* and the logic is called *epistemic logic*. (The most plausible epistemic logic is *T*.) It may be interpreted as "it is believed that," in which case it is usually written as *B*, and the logic is called *doxastic logic*. (Though even the logic *K* seems rather too strong here, except as an idealization to logically omniscient beings.) □ may be interpreted as "it is obligatory to bring it about that," in which case it is written as *O*, and the logic is called *deontic logic*. The standard deontic logic is *D*.

One can also interpret □ as "it is provable that." The best-known system in this regard is usually known as *GL* and called *provability logic*. This logic imposes just two constraints on the accessibility relation. One is transitivity; the other is that there are no infinite *R* -chains, that is, no sequences of the form w_0Rw_1, w_1Rw_2, w_2Rw_3, ... This constraint verifies the principle □(□*A* →*A*)→□*A*, but not □*A* →*A*. The interest of this system lies in its close connection with the way that a provability predicate, *Prov*, works in standard systems of formal arithmetic. By Gödel's second incompleteness theorem, in such logics one cannot prove *Prov* (⟨*A* ⟩) → *A* (where ⟨*A* ⟩ is the numeral for the gödel number of *A*); but Löb's theorem assures us that if we can prove *Prov*

$(\langle A \rangle) \to A$ we can prove A, and so *Prov* $(\langle A \rangle)$. It is this idea that is captured in the characteristic principle of *GL*.

Another possibility is to interpret □ and ◇ as, respectively, 'it will always be the case that,' and 'it will be the case at some time that.' In this context the operators are normally written as *G* and *F*, and the logic is called *tense logic*. In the world-semantics for tense logics, worlds are thought of as times, and the accessibility relation, *R*, is interpreted as a temporal ordering. In these logics there are also past-tense operators: *H* and *P* ("it has always been the case that" and "it was the case at some time that," respectively). These are given the reverse truth conditions. Thus for example:

$N_w(HA)=1$, *iff for all w' such that w'Rw*, $v_w(A)=1$

The past and future tense operators interact in characteristic ways (e.g., $A \to HFA$ is logically valid). The basic tense logic, K_t, is that obtained when *R* is arbitrary. As with modal logics, stronger systems are obtained by adding constraints on *R*, which can now represent the ideas that time is dense, has no last moment, and so on.

Of course it is not necessary to have just one family of intensional operators in a formal language: One can have, for example, modal and tense operators together. Each family will have its own accessibility relation, and these may interact in appropriate ways. Systems of logic with more than one family of modal operators are called *multi-modal*. One of the most important multi-modal logics is *dynamic logic*. In this there are operators of the form [α] and ⟨α⟩, each with its own accessibility relation, R_α. In the semantics of dynamic logic, the worlds are thought of as states of affairs or of a computational device. The αs are thought of as (non-deterministic) actions or programs, and $wR_\alpha w'$ is interpreted to mean that starting in state *w* and performing the action α (or running the program α) can take one to the state *w'*. Thus [α]A (⟨α⟩A) holds at state *w*, just if performing α at *w* will always (may sometimes) lead to a state in which *A* holds. The actions themselves are closed under certain operations. In particular, if α and β are actions, so are α;β (perform α and then perform β); α∪β (perform α or perform β, non-deterministically); α* (perform α some finite number of times, non-deterministically). There is also an operator,? ("test whether"), which takes sentences into programs. The corresponding accessibility relations are: $xR_{\alpha;\beta}y$ iff for some *z*, $xR_\alpha z$ and $zR_\beta y$; $xR_{\alpha\cup\beta}y$ iff $xR_\alpha y$ or $xR_\beta y$; $xR_{\alpha*}y$ iff for some $x=x_1, x_2, ..., x_n=y$, $x_0R_\alpha x_1, x_1R_\alpha x_2, ..., x_{n-1}R_\alpha x_n$; $xR_{A?}y$ iff ($x=y$ and $v_x(A)=1$). Because of the * operator, dynamic logic can express the notion of finitude in a certain sense. This gives it some of the expressive strength of second-order logic.

Conditional Logics

Another family of logics of the intentional variety was triggered by some apparent counter-examples to the following inferences:

$A \to B \vdash (A \wedge C) \to B$

$A \to B, B \to C \vdash A \to C$

$A \to B \vdash \neg B \to \neg A$

Which are valid for the material conditional. (For example: "If you strike this match it will light;

hence if you strike this match and it is under water it will light.") Logics of the conditional that invalidate such principles are called *conditional logics*. Such logics add an intentional conditional operator, >, to the language. In the semantics there is an accessibility relation, R_A, for every sentence, A (or one, R_x, for every proposition, that is, set of worlds, X). Intuitively wR_Aw' iff w' is a world which A holds but is, *ceteris paribus*, the same as w. The truth conditions for > are:

$N_w (A > B) = 1$ *iff for all w' such that $wR_A w'$, $v_{w'}(B) = 1$*

The intuitive meaning of R motivates the following constraints:

$WR_A w'$ *then $v_w(A) = 1$*

If $v_w (A) = 1$, then wR_Aw

Stronger logics in the family are obtained by adding further constraints to the accessibility relations. A standard way of specifying these is in terms of "similarity spheres"—neighbourhoods of a world containing those worlds that have a certain degree of similarity to it.

The natural way of taking a conditional logic is as a rival to classical logic (giving a different account of the conditional). Some philosophers, however, distinguish between indicative conditionals and subjunctive/counterfactual conditionals. They take the indicative conditional to be the material conditional of classical logic, and > to be the subjunctive conditional. Looked at this way conditional logics can be thought of as extensions of classical logic.

Intuitionist Logic

There are a number of other important non-classical logics that, though not presented originally as intentional logics, can be given world semantics. One of these is *intuitionist logic*. This logic arose out of a critique of Platonism in the philosophy of mathematics. The idea is that one cannot define truth in mathematics in terms of correspondence with some objective realm, as in a traditional approach. Rather one has to define it in terms of what can be proved, where a proof is something that one can effectively recognize as such. Thus, semantically, one has to replace standard truth-conditions with proof-conditions, of the following kind:

$A \lor B$ *is provable when A is provable or B is provable.*

$\neg A$ *is provable when it is provable that there is no proof of A*

$\exists x A (x)$ *is provable when we can effectively find an object, n, such that A(n) is provable*

Note that in the case of negation we cannot say that $\neg A$ is provable when A is not provable: We have no effective way of recognizing what is not provable; similarly, in the case of the existential quantifier, we cannot say that $\exists x A (x)$ is provable when there is some n such that $A(n)$ is provable: we may have no effective way of knowing whether this obtains.

Proceeding in this way produces a logic that invalidates a number of the principles of inference that are valid in classical logic. Notable examples are: $A \lor \neg A$, $\neg\neg A \to A$, $\neg\forall x A (x) \to \exists x \neg A (x)$. For the first of these, there is no reason to suppose that for any A we can find a proof of A or a proof that there is no proof of A. For the last, the fact that we can show that there is no proof of $\forall x A (x)$ does not mean that we can effectively find an n such that $A(n)$ can be proved.

In the world-semantics for intuitionist logic, interpretations have essentially the structure of an S 4 interpretation. The worlds are interpreted as states of information (things proved), and the accessibility relation represents the acquisition of new proofs. We also require that if $v_w(A)=1$ and wRw', $v_{w'}(A)=1$ (no information is lost), and if x is in the domain of quantification of w and wRw' then x is in the domain of quantification of w' (no objects are undiscovered). Corresponding to the provability conditions we have:

$N_w (A \vee B)=1$ iff $v_w(A)=1$ or $v_w(A)=1$

$N_w (\neg A)=1$ iff for all w' such that wRw', $v_{w'}(A)=0$

$N_w (\exists x A (x))=1$ iff for all n in the domain of w, $v_w (A (n))=1$

Unsurprisingly, given the above semantics, there is a translation of the language of intuitionism into quantified S 4 that preserves validity.

Another sort of semantics for intuitionism takes semantic values to be the open sets of some topology. If the value of A is x, the value of $\neg A$ is the interior of the complement of x.

Relevant Logic

Another logic standardly thought of as a rival to classical logic is *relevant* (or *relevance*) logic. This is motivated by the apparent incorrectness of classical validities such as: $A \rightarrow (B \rightarrow B)$, $(A \wedge \neg A) \rightarrow B$. A (propositional) relevant logic in one in which if $A \rightarrow B$ is a logical truth A and B share a propositional parameter. There are a number of different kinds of relevant logic, but the most common has a world-semantics. The semantics differs in two major ways from the world semantics we have so far met.

First it adds to the possible worlds a class of logically impossible worlds. (Though validity is still defined in terms of truth-preservation over possible worlds.) In possible worlds the truth conditions of $\rightarrow$ are as for $\dashv$ in S 5:

$N_w (A \rightarrow B)=1$ iff for all w' (possible and impossible) such that $v_w(A)=1$, $v_w(B)=1$

In impossible worlds the truth conditions are given differently, in such a way that logical laws such as $B \rightarrow B$ may fail at the world. This may be done in various ways, but the most versatile technique employs a three-place relation, S, on worlds. If w is impossible, we then have:

$N_w (A \rightarrow B)=1$ iff for all x,y such that $Swxy$, if $v_x (A)=1$, $v_y (B)=1$

This clause can be taken to state the truth conditions of $\rightarrow$ at all worlds, provided that we add the constraint that, for possible w, $Swxy$ iff $x =y$. With no other constraints on S, this gives the basic (positive) relevant logic, B. Additional constraints on S give stronger logics in the family. Typical constraints are:

$\exists x (Sabx$ and $Sxcd) \Rightarrow \exists y (Sacy$ and $Sbyd)$

$Sabc \Rightarrow Sbac$

$Sabc \Rightarrow \exists x (Sabx$ and $Sxbc)$

Adding all three gives the (positive) relevant logic, R. Adding the first two gives RW, R minus Contraction ($A \to (A \to B) \vdash A \to B$). The intuitive meaning of S is, at the time of this writing, philosophically moot.

The second novelty of the semantics is in its treatment of negation. It is necessary to arrange for worlds where $A \wedge \neg A$ may hold. This may be done in a couple of ways. The first is to employ the Routley * operator. Each world, w, comes with a "mate," w^* (subject to the constraint that $w^{**}=w$, to give Double Negation). We then have:

$$N_w (\neg A) = 1 \text{ iff } v_{w^*}(A) = 0$$

(If $w = w^*$, this just delivers the classical truth conditions.) Alternatively, we may move to a four-valued logic in which the values at a world are *true only, false only, both, neither* ({1}, {0}, {1,0}, $\emptyset$). We then have:

$$1 \in v_w (\neg A) \text{ iff } 0 \in v_w (A)$$

$$0 \in v_w (\neg A) \text{ iff } 1 \in v_w (A)$$

The semantics of relevant logic can be extended to produce a (relevant) *ceteris paribus* conditional, >, of the kind found in conditional logics, by adding the appropriate binary accessibility relations.

Distribution-free Logics

There are some logics in the family of relevant logics for which the principle of Distribution, $A \wedge (B \vee C) \vdash (A \wedge B) \vee (A \wedge C)$, fails. To achieve this the truth conditions for disjunction have to be changed. In an interpretation, let $[A]$ be the set of worlds at which A holds. Then the usual truth conditions for disjunction can be written:

$$N_w (A \vee B) = 1 \text{ iff } w \in [A] \cup [B]$$

To invalidate Distribution, the semantics are augmented by a closure operator, $\mathfrak{C}$, on sets of worlds, x, satisfying the following conditions:

$$X \subseteq \mathfrak{C}(X)$$

$$\mathfrak{C}\mathfrak{C}(X) = \mathfrak{C}X$$

$$\text{If } X \subseteq Y \text{ then } \mathfrak{C}(X) \subseteq \mathfrak{C}(Y)$$

The truth conditions of disjunction can now be given as:

$$N_w (A \vee B) = 1 \text{ iff } w \in \mathfrak{C}([A] \cup [B])$$

Changing the truth conditions for disjunction in RW in this way (and using the Routley * for negation) gives linear logic (LL). LL is usually formulated with some extra intentional connectives, especially an intentional conjunction and disjunction. These connectives can be present in standard relevant logics too. Intuitionist, relevant, and linear logics all belong to the family of *substructural logics*. Proof-theoretically, these logics can be obtained from a sequent-calculus for classical logic by weakening the structural rules (especially Weakening and Contraction).

Another logic in which distribution fails is *quantum logic*. The thought here is that it may be true (verifiable) of a particle that it has a position and one of a range of momenta, but each disjunct attributing to it that position and a particular momentum is false (unverifiable). The states of a quantum system are canonically thought of as members of a Hilbert space. In the world-semantics for quantum logic, the space of worlds is taken to be such a space, and sentences are assigned closed subsets of this. $[A \wedge B] = [A] \cap [B]$, $[A \vee B] = \mathfrak{C}([A] \cup [B])$, where $\mathfrak{C}(X)$ is the smallest closed space containing X; and $[\neg A] = [A]^{\perp}$. $X^{\perp}$ is the space comprising all those states that are orthogonal to members of X. (It satisfies the conditions: $X = X^{\perp\perp}$, if $X \subseteq Y$ then $Y^{\perp} \subseteq X^{\perp}$, and $X \cap X^{\perp} = \emptyset$.) In quantum logic $A \rightarrow B$ can be defined in various ways. Perhaps the most plausible is as $\neg A \vee (A \wedge B)$. (The subspaces of a Hilbert space also have the structure of a partial Boolean algebra. Such an algebra is determined by a family of Boolean algebras collapsed under a certain equivalence relation, which is a congruence relation on the Boolean operators. Partial Boolean algebras can be used to provide a slightly different quantum logic.)

Paraconsistent Logics

Before we turn to quantifiers there is one further kind of logic to be mentioned: *paraconsistent logic*. Paraconsistent logic is motivated by the thought we would often seem to have to reason sensibly from information, or about a situation, which is inconsistent. In such a case, the principle $A, \neg A \vdash B$ (*ex falso quodlibet sequitur*, Explosion), which is valid in classical logic, clearly makes a mess of things. A paraconsistent logic is precisely one where this principle fails.

There are many different families of paraconsistent logics—as many as there are ways of breaking Explosion. Indeed many of the techniques we have already met in this article can be used to construct a paraconsistent logic. The 3-valued logic *LP* is paraconsistent, as is the Łukasiewicz continuum-valued logic, provided we take the designated values to contain 0.5. The ways that negation is handled in relevant logic also produce paraconsistent logics, as long as validity is defined over a class of worlds in which A and ¬A may both hold. Another approach (*discussive logic*) is to employ standard modal logic and to take A to hold in an interpretation iff A holds at some world of the interpretation. In this approach the principle of Adjunction ($A, B \vdash A \wedge B$) will generally fail, since A and B may each hold at a world, whilst $A \wedge B$ may not. Another approach ("positive plus") is to take any standard positive (negation free) logic, and add a non-truth-functional negation—so that the values of A and ¬A are assigned independently. In these logics, the principle of Contraposition ($A \leftrightarrow B \vdash \neg B \leftrightarrow \neg A$) will generally fail. Yet another is to dualise intuitionist logic. In particular one can take semantic values to be the closed sets in some topology. If the value of A is X, the value of ¬A is the closure of the complement of X.

Second-order Quantification

We now turn to the issue of quantification. In classical logic there are quantifiers $\forall$ and $\exists$. These range over a domain of objects, and $\forall x A(x)$ [$\exists x A(x)$] holds if every [some] object in the domain of quantification satisfies $A(x)$. All the propositional logics we have looked at may be extended to first-order logics with such quantifiers. Other non-classical logics may be obtained by adding to these (or replacing these with) different kinds of quantifiers.

Perhaps the most notable of these is second-order logic. In this there are bindable variables (*X, Y, ...*) that can stand in the place where a monadic first-order predicate can stand and which range over sets of objects in the first-order domain—canonically all of them. (There can also be variables that range over the n-ary relations on that domain, for each n, as well as variables that range over n-place functions. The second-order extension of classical logic is much stronger than the first-order version. It can provide for a categorical axiomatization of arithmetic and consequently is not itself axiomatizable.

Monadic second-order quantifiers can also be given a rather different interpretation, as plural quantifiers. The idea here is to interpret $\exists X\,Xa$ not as "There is a set such that a is a member of it," but as "There are some things such that a is one of them." The proponents of plural quantification argue that such quantification is not committed to the existence of sets.

Other Sorts of Quantifiers

There are many other non-classical quantifiers. For example one can have a binary quantifier of the form $Mx\,(A\,(x\,),B\,(x\,))$, "most A s are B s." This is true in a finite domain if more than half the things satisfying $A\,(x)$ satisfy $B\,(x\,)$. It is not reducible to a monadic quantifier plus a propositional connective.

Another sort of quantifier is a cardinality quantifier. The quantifier "there exist exactly n things such that" can be defined in first-order logic with quantification and identity in a standard way. The quantifier "there is a countable number of things such that" (or its negation, "there is an uncountable number of things such that") cannot be so defined—let alone the quantifier "there are κ things such that," for an arbitrary cardinal, κ. Such quantifiers can be added, with the obvious semantics. These quantifiers extend the expressive power of the language towards that of second-order logic—and beyond.

Another kind of quantifier is the branching quantifier. When, in first-order logic, we write:

$$\forall x_1\,\exists y_1\,\forall x_2\,\exists y_2\,A\,(x_1,x_2,y_1,y_2)$$

Y_2 is in the scope of x_1, and so its value depends on that of x_1. To express non-dependence one would normally need second-order quantification, thus:

$$\exists f_1\,\forall x_1\,\exists f_2\,\forall x_2\,A\,(x_1,x_2,f_1(x_1),f_2(x_2))$$

But we may express it equally by having the quantifiers non-linearly ordered, thus:

As this would suggest, branching quantifiers have something of the power of second-order logic.

A quite different kind of quantifier is the substitutional quantifier. For this there is a certain class of names of the language, C. $\Pi x A\,(x\,)$ [$\Sigma x A\,(x\,)$] holds iff for every [some] $c \in C, A\,(c\,)$ holds. This is not the same as standard (objectual) quantification, since some objects in the domain may have no name in C; but first-order substitutional quantifiers validate the same quantificational inferences as first-order objectual quantifiers. Note that the notion of substitutional quantification makes perfectly good sense for any syntactically well-defined class, including predicates (so we can have second-order substitutional quantification) or binary connectives (so that $\Sigma x\,(AxB\,)$ can make perfectly good sense).

Finally in this category comes free quantifiers. It is standard to interpret the domain of objects of quantification (at a world) as comprising the objects that exist (at that world). It is quite possible, however, to think of the domain as containing a bunch of objects, some of which exist, and some of which do not. Obviously this does not change the formal properties of the quantifiers. But if one thinks of the domain in this way one must obviously not read $\exists x$ as 'there exists an x such that'; one has to read it simply as 'for some x'. Given this set-up, however, it makes sense to have existentially loaded quantifiers, $\forall^E$ and $\exists^E$, such that $\forall^E A (x)$ [$\exists^E A (x)$] holds (at a world) iff all [some] of the existent objects (at the world) satisfy $A (x)$. If there is a monadic existence predicate, E, these quantifiers can be defined in the obvious way, as (respectively): $\forall x (Ex \rightarrow A (x))$ and $\exists x (Ex \wedge A (x))$. Clearly, existentially loaded quantifiers will not satisfy some of the standard principles of quantification, such as $\forall^E x A (x) \rightarrow A (c)$, $A (c) \rightarrow \exists x^E A (x)$ (since the object denoted by 'c' may not exist). Some logics do not have the existentially unloaded quantifiers, just the loaded ones. These are usually called *free logics*.

Non-monotonic Logics

It remains to say a word about one other kind of logic that is often categorized as non-classical. In all the logics we have been considering so far:

> *If $\Sigma \vdash A$ then $\Sigma \cup \Delta \vdash A$*

(Where Σ and Δ are sets of formulas): Adding extra premises makes no difference. This is called *monotonicity*. Logics in which this principle fails are called *non-monotonic logics*. Non-monotonic inferences can be thought of as inferences that are made with certain default assumptions. Thus I am told that something is a bird, and I infer that it can fly. Since most birds fly this is a reasonable conclusion. If, however, I also learn that the bird weighs 20 kg. (and so is an emu or an ostrich), the conclusion is no longer a reasonable one.

There are many kinds of non-monotonic logics, depending on what kind of default assumption is implemented, but there is a common structure that covers many of them. Interpretations, I, of the language come with a strict partial ordering, $>$ (often called a *preference ordering*). Intuitively, $I_1 > I_2$ means that the situation represented by I_1 is more normal (in whatever sense of normality is at issue) than that represented by I_2. (In particular cases it may be reasonable to suppose that $>$ has additional properties.) I is a *most normal model of* Σ iff every $B \in \Sigma$ holds in I, and there is no $J > I$ for which this is true. A follows from Σ iff A holds in every most normal model of Σ. As is clear a most normal model of Σ is not guaranteed to be a most normal model of $\Sigma \cup \Delta$. Hence monotonicity will fail. As might be expected there is a close connection between non-monotonic logics and conditional logics, in which the inference $A \rightarrow B \vdash (A \wedge C) \rightarrow B$ fails. Though non-monotonic logic has come to prominence in modern computational logic, it is just a novel and rigorous way of looking at the very traditional notion of non-deductive (inductive, ampliative) inference.

History, Persons, References

We conclude this review of non-classical logics by putting the investigations discussed above in their historical context. References that may be consulted for further details are also given at the end of each paragraph. For a general introduction to propositional non-classical logics. Haack (1996) is a discussion of some of the philosophical issues raised by non-classical logics.

The first modern many-valued logics, the $Ł_n$ family, were produced by Jan Łukasiewicz in the early 1920s. (Emil Post also produced some many-valued logics about the same time.) Łukasiewicz's major philosophical concern was Aristotle's argument for fatalism. In this context he suggested a many-valued analysis of modality. Logics of the both/neither kind were developed somewhat later. Canonical statements of K_3 and *LP* were given (respectively) by Stephen Kleene in the 1950s and Graham Priest in the 1970s. $Ł_\aleph$ was first published by Łukasiewicz and Alfred Tarski in 1930. The intensive investigation of fuzzy logics and their applications started in the 1970s. A notable player in this area was Lotfi Zadeh.

Modern modal logics were created in an axiomatic form by Clarence Irving Lewis in the 1920s. Lewis's concern was the paradoxes of the material conditional, and he suggested the strict conditional as an improvement. Possible-world semantics for modal logics were produced by a number of people in the 1960s, but principally Saul Kripke. The semantics made possible the systematic investigation of the rich family of modal logics.

The idea that the techniques of modal logics could be applied to notions other than necessity and possibility occurred to a number of people around the middle of the twentieth century. Tense logics were created by Arthur Prior, epistemic and doxastic logic were produced by Jaakko Hintikka, and deontic logics by Henrik von Wright. Investigations of provability logic were started in the 1970s by George Boolos and others. Dynamic logic was created by Vaughn Pratt and other logicians particularly interested in computation, including David Harrel, in the 1970s.

Conditional logics (with "sphere semantics") were proposed by David Lewis and Robert Stalnaker in the 1970s. They were formulated as multi-modal logics by Brian Chellas and Krister Segerberg a few years later.

The intuitionist critique of classical mathematics was started by Luitzen Egbertus Jan Brouwer in the early years of the twentieth century. This generated a novel kind of mathematics: intuitionist mathematics. Intuitionist logic, as such, was formulated by Arend Heyting and Andrei Kolmogorov in the 1920s. The intuitionist critique of mathematical realism was extended to realism in general by Michael Dummett in the 1970s.

Systems of relevant logic, in axiomatic form, came to prominence in the 1960s because of the work of Alan Anderson, Nuel Belnap and their students. World-semantics were produced by a number of people in the 1970s, but principally Richard Routley (later Sylvan) and Robert Meyer. The semantics made possible the investigation of the rich family of relevant logics. The four-valued semantics for negation is due to J. Michael Dunn.

Linear logic was produced by Jean-Yves Girard in the 1980s. Although many members of the class of sub-structural logics had been studied before, the fact that they could be viewed in a uniform proof-theoretic way, was not appreciated until the late 1980s. The formulation of quantum logic in terms of Hilbert spaces is due, essentially, to George Birkhoff and John von Neumann in the 1930s. The use of an abstract closure operator to give the semantics for non-distributive logics is due to Greg Restall.

The first paraconsistent logic (discussive logic) was published by StanisŁaw Jaśkowski in 1948. Other non-adjunctive logics were later developed in the 1970s by Peter Schotch and Raymond Jennings. Newton da Costa produced a number of different paraconsistent logics and applications, starting with positive-plus logics in the 1960s. The paraconsistent aspects of relevant logic were developed by Priest and Routley in the 1970s.

Second-order quantification goes back to the origins of classical logic in the work of Gottlob Frege and Bertrand Russell. Its unaxiomatizability put it somewhat out of fashion for a number of years, but it made a strong come-back in the last years of the twentieth century. The notion of plural quantification was made popular by George Boolos in the 1980s.

Quantifier phrases other than "some A " and "all A " are pervasive in natural language; and since Frege provided an analysis of the quantifier many different kinds have been investigated by linguists and logicians. Branching quantifiers were proposed by Jaakko Hintikka in the 1970s. Substitutional quantification came to prominence in the 1960s, put there particularly in connection with quantification into the scope of modal operators by Ruth Barcan Marcus. It was treated with suspicion for a long time, but was eventually given a clean bill of health by Kripke. Free logics were first proposed in the 1960s, by Karel Lambert and others.

Non-monotonic logics started to appear in the logic/computer-science literature in the 1970s. There are many kinds. The fact that many of them could be seen as logics with normality orderings started to become clear in the 1980s.

Modal Logics

The most straightforward way of constructing a modal logic is to add to some standard nonmodal logical system a new primitive operator intended to represent one of the modalities, to define other modal operators in terms of it, and to add axioms or transformation rules involving those modal operators. For example, one may add the symbol L, which means "It is necessary that," to the classical propositional calculus; thus, Lp is read as "It is necessary that p." The possibility operator M ("It is possible that") may be defined in terms of L as Mp = ¬L¬p (where ¬ means "not"). In addition to the axioms and rules of inference of classical propositional logic, such a system might have two axioms and one rule of inference of its own. Some characteristic axioms of modal logic are: Lp ⊃ p and L(p ⊃ q) ⊃ (Lp ⊃ Lq). The new rule of inference in this system is the rule of necessitation: if p is a theorem of the system, then so is Lp. Stronger systems of modal logic can be obtained by adding additional axioms. For example, some add the axiom Lp ⊃ LLp, while others add the axiom Mp ⊃ LMp.

Deontic Logics

Deontic logics introduce the primitive symbol O for 'it is obligatory that', from which symbols P for 'it is permitted that' and F for 'it is forbidden that' are defined: PA = ~O~A and FA = O~A. The deontic analog of the modal axiom (M): OA→A is clearly not appropriate for deontic logic. (Unfortunately, what ought to be is not always the case.) However, a basic system D of deontic logic can be constructed by adding the weaker axiom (D) to K.

> (D) OA→PA

Axiom (D) guarantees the consistency of the system of obligations by insisting that when A is obligatory, A is permissible. A system which obligates us to bring about A, but doesn't permit us to do so, puts us in an inescapable bind. Although some will argue that such conflicts of obligation are at least possible, most deontic logicians accept (D).

O(OA→A) is another deontic axiom that seems desirable. Although it is wrong to say that if A is obligatory then A is the case (OA→A), still, this conditional ought to be the case. So some deontic logicians believe that D needs to be supplemented with O(OA→A) as well.

Controversy about iteration (repetition) of operators arises again in deontic logic. In some conceptions of obligation, OOA just amounts to OA. 'It ought to be that it ought to be' is treated as a sort of stuttering; the extra 'ought's do not add anything new. So axioms are added to guarantee the equivalence of OOA and OA. The more general iteration policy embodied in S5 may also be adopted. However, there are conceptions of obligation where distinction between OA and OOA is preserved. The idea is that there are genuine differences between the obligations we actually have and the obligations we should adopt. So, for example, 'it ought to be that it ought to be that A' commands adoption of some obligation which may not actually be in place, with the result that OOA can be true even when OA is false.

Temporal Logics

In temporal logic (also known as tense logic), there are two basic operators, G for the future, and H for the past. G is read 'it always will be that' and the defined operator F (read 'it will be the case that') can be introduced by FA = ~G~A. Similarly H is read: 'it always was that' and P (for 'it was the case that') is defined by PA=~H~A. A basic system of temporal logic called Kt results from adopting the principles of K for both G and H, along with two axioms to govern the interaction between the past and future operators:

"Necessitation" Rules:

>   If A is a theorem then so are GA and HA.

Distribution Axioms:

>   G(A→B) → (GA→GB) and H(A→B) → (HA→HB)

Interaction Axioms:

>   A→GPA and A→HFA

The interaction axioms raise questions concerning asymmetries between the past and the future. A standard intuition is that the past is fixed, while the future is still open. The first interaction axiom (A→GPA) conforms to this intuition in reporting that what is the case (A), will at all future times, be in the past (GPA). However A→HFA may appear to have unacceptably deterministic overtones, for it claims, apparently, that what is true now (A) has always been such that it will occur in the future (HFA). However, possible world semantics for temporal logic reveals that this worry results from a simple confusion, and that the two interaction axioms are equally acceptable.

Note that the characteristic axiom of modal logic, (M): □A→A, is not acceptable for either H or G, since A does not follow from 'it always was the case that A', nor from 'it always will be the case that A'. However, it is acceptable in a closely related temporal logic where G is read 'it is and always will be', and H is read 'it is and always was'.

Depending on which assumptions one makes about the structure of time, further axioms must be added to temporal logics. A list of axioms commonly adopted in temporal logics follows. An account of how they depend on the structure of time will be found in the section Possible Worlds Semantics.

GA→GGA and HA→HHA

GGA→GA and HHA→HA

GA→FA and HA→PA

It is interesting to note that certain combinations of past tense and future tense operators may be used to express complex tenses in English. For example, FPA, corresponds to sentence A in the future perfect tense, (as in '20 seconds from now the light will have changed'). Similarly, PPA expresses the past perfect tense.

Conditional and Relevance Logics

The founder of modal logic, C. I. Lewis, defined a series of modal logics which did not have □ as a primitive symbol. Lewis was concerned to develop a logic of conditionals that was free of the so called Paradoxes of Material Implication, namely the classical theorems A→(~A→B) and B→(A→B). He introduced the symbol -3 for "strict implication" and developed logics where neither A -3 (~A -3 B) nor B -3 (A -3 B) is provable. The modern practice has been to define A -3 B by □(A→B), and use modal logics governing □ to obtain similar results. However, the provability of such formulas as (A&~A) -3 B in such logics seems at odds with concern for the paradoxes. Anderson and Belnap (1975) have developed systems R (for Relevance Logic) and E (for Entailment) which are designed to overcome such difficulties. These systems require revision of the standard systems of propositional logic.

David Lewis (1973) and others have developed conditional logics to handle counterfactual expressions, that is, expressions of the form 'if A were to happen then B would happen'. (Kvart (1980) is another good source on the topic.) Counterfactual logics differ from those based on strict implication because the former reject while the latter accept contraposition.

Possible Worlds Semantics

The purpose of logic is to characterize the difference between valid and invalid arguments. A logical system for a language is a set of axioms and rules designed to prove exactly the valid arguments statable in the language. Creating such a logic may be a difficult task. The logician must make sure that the system is sound, i.e. that every argument proven using the rules and axioms is in fact valid. Furthermore, the system should be complete, meaning that every valid argument has a proof in the system. Demonstrating soundness and completeness of formal systems is a logician's central concern.

Such a demonstration cannot get underway until the concept of validity is defined rigorously. Formal semantics for a logic provides a definition of validity by characterizing the truth behavior of the sentences of the system. In propositional logic, validity can be defined using truth tables. A valid argument is simply one where every truth table row that makes its premises true also makes its conclusion true. However truth tables cannot be used to provide an account of validity in modal logics because there are no truth tables for expressions such as 'it is necessary that', 'it is obligatory that', and the like. (The problem is that the truth value of A does not determine the truth value for

□A. For example, when A is 'Dogs are dogs', □A is true, but when A is 'Dogs are pets', □A is false.) Nevertheless, semantics for modal logics can be defined by introducing possible worlds. We will illustrate possible worlds semantics for a logic of necessity containing the symbols ~, →, and □. Then we will explain how the same strategy may be adapted to other logics in the modal family.

In propositional logic, a valuation of the atomic sentences (or row of a truth table) assigns a truth value (T or F) to each propositional variable p. Then the truth values of the complex sentences are calculated with truth tables. In modal semantics, a set W of possible worlds is introduced. A valuation then gives a truth value to each propositional variable for each of the possible worlds in W. This means that value assigned to p for world w may differ from the value assigned to p for another world w'.

The truth value of the atomic sentence p at world w given by the valuation v may be written v(p, w). Given this notation, the truth values (T for true, F for false) of complex sentences of modal logic for a given valuation v (and member w of the set of worlds W) may be defined by the following truth clauses. ('iff' abbreviates 'if and only if'.)

(~) v(~A, w)=T iff v(A, w)=F.

(→) v(A→B, w)=T iff v(A, w)=F or v(B, w)=T.

(5) v(□A, w)=T iff for every world w' in W, v(A, w')=T.

Clauses (~) and (→) simply describe the standard truth table behavior for negation and material implication respectively. According to (5), □A is true (at a world w) exactly when A is true in all possible worlds. Given the definition of ◊, (namely, ◊A = ~□~A) the truth condition (5) insures that ◊A is true just in case A is true in some possible world. Since the truth clauses for □ and ◊ involve the quantifiers 'all' and 'some' (respectively), the parallels in logical behavior between □ and ∀x, and between ◊ and ∃x.

Clauses (~), (→), and (5) allow us to calculate the truth value of any sentence at any world on a given valuation. A definition of validity is now just around the corner. An argument is 5-valid for a given set W (of possible worlds) if and only if every valuation of the atomic sentences that assigns the premises T at a world in W also assigns the conclusion T at the same world. An argument is said to be 5-valid iff it is valid for every non empty set W of possible worlds.

It has been shown that S5 is sound and complete for 5-validity (hence our use of the symbol '5'). The 5-valid arguments are exactly the arguments provable in S5. This result suggests that S5 is the correct way to formulate a logic of necessity.

However, S5 is not a reasonable logic for all members of the modal family. In deontic logic, temporal logic, and others, the analog of the truth condition (5) is clearly not appropriate; furthermore there are even conceptions of necessity where (5) should be rejected as well. The point is easiest to see in the case of temporal logic. Here, the members of W are moments of time, or worlds "frozen", as it were, at an instant. For simplicity let us consider a future temporal logic, a logic where □A reads: 'it will always be the case that'. (We formulate the system using □ rather than the traditional G so that the connections with other modal logics will be easier to appreciate.) The correct clause for □ should say that □A is true at time w iff A is true at all times in the future of w. To restrict attention to the future, the relation R (for 'eaRlier than') needs to be introduced. Then the correct clause can be formulated as follows.

(K) $v(\Box A, w)=T$ iff for every w', if wRw', then $v(A, w')=T$.

This says that $\Box A$ is true at w just in case A is true at all times after w.

Validity for this brand of temporal logic can now be defined. A frame $<W, R>$ is a pair consisting of a non-empty set W (of worlds) and a binary relation R on W. A model $<F, v>$ consists of a frame F, and a valuation v that assigns truth values to each atomic sentence at each world in W. Given a model, the values of all complex sentences can be determined using $(\sim)$, $(\rightarrow)$, and (K). An argument is K-valid just in case any model whose valuation assigns the premises T at a world also assigns the conclusion T at the same world. As the reader may have guessed from our use of 'K', it has been shown that the simplest modal logic K is both sound and complete for K-validity.

Modal Axioms and Conditions on Frames

One might assume from this discussion that K is the correct logic when $\Box$ is read 'it will always be the case that'. However, there are reasons for thinking that K is too weak. One obvious logical feature of the relation R (earlier than) is transitivity. If wRv (w is earlier than v) and vRu (v is earlier than u), then it follows that wRu (w is earlier than u). So let us define a new kind of validity that corresponds to this condition on R. Let a 4-model be any model whose frame $<W, R>$ is such that R is a transitive relation on W. Then an argument is 4-valid iff any 4-model whose valuation assigns T to the premises at a world also assigns T to the conclusion at the same world. We use '4' to describe such a transitive model because the logic which is adequate (both sound and complete) for 4-validity is K4, the logic which results from adding the axiom (4): $\Box A \rightarrow \Box \Box A$ to K.

Transitivity is not the only property which we might want to require of the frame $<W, R>$ if R is to be read 'earlier than' and W is a set of moments. One condition (which is only mildly controversial) is that there is no last moment of time, i.e. that for every world w there is some world v such that wRv. This condition on frames is called seriality. Seriality corresponds to the axiom (D): $\Box A \rightarrow \Diamond A$, in the same way that transitivity corresponds to (4). A D-model is a K-model with a serial frame. From the concept of a D-model the corresponding notion of D-validity can be defined just as we did in the case of 4-validity. As you probably guessed, the system that is adequate with respect to D-validity is KD, or K plus (D). Not only that, but the system KD4 (that is K plus (4) and (D)) is adequate with respect to D4-validity, where a D4-model is one where $<W, R>$ is both serial and transitive.

Another property which we might want for the relation 'earlier than' is density, the condition which says that between any two times we can always find another. Density would be false if time were atomic, i.e. if there were intervals of time which could not be broken down into any smaller parts. Density corresponds to the axiom (C4): $\Box \Box A \rightarrow \Box A$, the converse of (4), so for example, the system KC4, which is K plus (C4) is adequate with respect to models where the frame $<W, R>$ is dense, and KDC4, adequate with respect to models whose frames are serial and dense, and so on.

Each of the modal logic axioms we have discussed corresponds to a condition on frames in the same way. The relationship between conditions on frames and corresponding axioms is one of the central topics in the study of modal logics. Once an interpretation of the intensional operator $\Box$ has been decided on, the appropriate conditions on R can be determined to fix the corresponding notion of validity. This, in turn, allows us to select the right set of axioms for that logic.

For example, consider a deontic logic, where □ is read 'it is obligatory that'. Here the truth of □A does not demand the truth of A in every possible world, but only in a subset of those worlds where people do what they ought. So we will want to introduce a relation R for this kind of logic as well, and use the truth clause (K) to evaluate □A at a world. However, in this case, R is not earlier than. Instead wRw' holds just in case world w' is a morally acceptable variant of w, i.e. a world that our actions can bring about which satisfies what is morally correct, or right, or just. Under such a reading, it should be clear that the relevant frames should obey seriality, the condition that requires that each possible world have a morally acceptable variant. The analysis of the properties desired for R makes it clear that a basic deontic logic can be formulated by adding the axiom (D) and to K.

Even in modal logic, one may wish to restrict the range of possible worlds which are relevant in determining whether □A is true at a given world. For example, I might say that it is necessary for me to pay my bills, even though I know full well that there is a possible world where I fail to pay them. In ordinary speech, the claim that A is necessary does not require the truth of A in all possible worlds, but rather only in a certain class of worlds which I have in mind (for example, worlds where I avoid penalties for failure to pay). In order to provide a generic treatment of necessity, we must say that □A is true in w iff A is true in all worlds that are related to w in the right way. So for an operator □ interpreted as necessity, we introduce a corresponding relation R on the set of possible worlds W, traditionally called the accessibility relation. The accessibility relation R holds between worlds w and w' iff w' is possible given the facts of w. Under this reading for R, it should be clear that frames for modal logic should be reflexive. It follows that modal logics should be founded on M, the system that results from adding (M) to K. Depending on exactly how the accessibility relation is understood, symmetry and transitivity may also be desired.

A list of some of the more commonly discussed conditions on frames and their corresponding axioms along with a map showing the relationship between the various modal logics can be found.

Map of the Relationships between Modal Logics

The following diagram shows the relationships between the best known modal logics, namely logics that can be formed by adding a selection of the axioms (D), (M), (4), (B) and (5) to K. A list of these (and other) axioms along with their corresponding frame conditions can be found below the diagram.

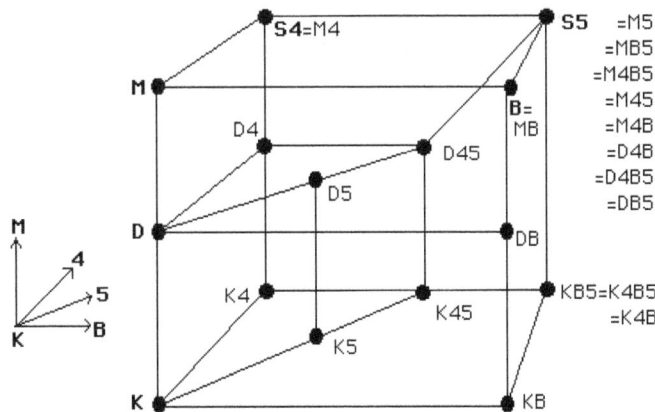

In this chart, systems are given by the list of their axioms. So, for example M4B is the result of adding (M), (4) and (B) to K. In boldface, we have indicated traditional names of some systems. When system S appears below and/or to the left of S' connected by a line, then S' is an extension of S. This means that every argument provable in S is provable in S', but S is weaker than S', i.e. not all arguments provable in S' are provable in S.

The following list indicates axioms, their names, and the corresponding conditions on the accessibility relation R, for axioms so far discussed in this table.

Axiom Name	Axiom	Condition on Frames	R is
(D)	$\Box A \rightarrow \Diamond A$	$\exists u\ wRu$	Serial
(M)	$\Box A \rightarrow A$	wRw	Reflexive
(4)	$\Box A \rightarrow \Box\Box A$	$(wRv\ \&\ vRu) \Rightarrow wRu$	Transitive
(B)	$A \rightarrow \Box\Diamond A$	$wRv \Rightarrow vRw$	Symmetric
(5)	$\Diamond A \rightarrow \Box\Diamond A$	$(wRv\ \&\ wRu) \Rightarrow vRu$	Euclidean
(CD)	$\Diamond A \rightarrow \Box A$	$(wRv\ \&\ wRu) \Rightarrow v=u$	Functional
$(\Box M)$	$\Box(\Box A \rightarrow A)$	$wRv \Rightarrow vRv$	Shift Reflexive
$(C4)$	$\Box\Box A \rightarrow \Box A$	$wRv \Rightarrow \exists u(wRu\ \&\ uRv)$	Dense
(C)	$\Diamond\Box A \rightarrow \Box\Diamond A$	$wRv\ \&\ wRx \Rightarrow \exists u(vRu\ \&\ xRu)$	Convergent

In the list of conditions on frames, and in the rest of this topic, the variables 'w', 'v', 'u', 'x' and the quantifier '$\exists u$' are understood to range over W. '&' abbreviates 'and' and '$\Rightarrow$' abbreviates 'if...then'.

The notion of correspondence between axioms and frame conditions that is at issue here was explained. When S is a list of axioms and F(S) is the corresponding set of frame conditions, then S corresponds to F(S) exactly when the system K+S is adequate (sound and complete) for F(S)-validity, that is, an argument is provable in K+S iff it is F(S)-valid. Several stronger notions of correspondence between axioms and frame conditions have emerged in research on modal logic.

The General Axiom

The correspondence between axioms and conditions on frames may seem something of a mystery. A beautiful result of Lemmon and Scott (1977) goes a long way towards explaining those relationships. Their theorem concerned axioms which have the following form:

(G) $\Diamond^h\Box^i A \rightarrow \Box^j\Diamond^k A$

We use the notation '$\Diamond^n$' to represent n diamonds in a row, so, for example, '$\Diamond^3$' abbreviates a string of three diamonds: '$\Diamond\Diamond\Diamond$'. Similarly '$\Box^n$' represents a string of n boxes. When the values of h, i, j, and k are all 1, we have axiom (C):

(C) $\Diamond\Box A \rightarrow \Box\Diamond A\ =\ \Diamond^1\Box^{1}A \rightarrow \Box^1\Diamond^{1}A$

The axiom (B) results from setting h and i to 0, and letting j and k be 1:

(B) $A \rightarrow \Box\Diamond A\ =\ \Diamond^0\Box^{0}A \rightarrow \Box^1\Diamond^{1}A$

To obtain (4), we may set h and k to 0, set i to 1 and j to 2:

(4) $\Box A \rightarrow \Box\Box A = \Diamond^0\Box^{1}A \rightarrow \Box^2\Diamond^0A$

Many (but not all) axioms of modal logic can be obtained by setting the right values for the parameters in (G)

Our next task will be to give the condition on frames which corresponds to (G) for a given selection of values for h, i, j, and k. In order to do so, we will need a definition. The composition of two relations R and R' is a new relation R ∘ R' which is defined as follows:

$wR \circ R'v$ iff for some u, wRu and $uR'v$.

For example, if R is the relation of being a brother, and R' is the relation of being a parent then $R \circ R'$ is the relation of being an uncle, (because w is the uncle of v iff for some person u, both w is the brother of u and u is the parent of v). A relation may be composed with itself. For example, when R is the relation of being a parent, then $R \circ R$ is the relation of being a grandparent, and $R \circ R \circ R$ is the relation of being a great-grandparent. It will be useful to write 'R^n', for the result of composing R with itself n times. So R^2 is $R \circ R$, and R^4 is $R \circ R \circ R \circ R$. We will let R^1 be R, and R^0 will be the identity relation, i.e. wR^0v iff $w=v$.

We may now state the Scott-Lemmon result. It is that the condition on frames which corresponds exactly to any axiom of the shape (G) is the following.

(hijk-Convergence) wR^hv & $wR^ju \Rightarrow \exists x (vR^ix$ & $uR^kx)$

It is interesting to see how the familiar conditions on R result from setting the values for h, i, j, and k according to the values in the corresponding axiom. For example, consider (5). In this case $i=0$, and $h=j=k=1$. So the corresponding condition is

wRv & $wRu \Rightarrow \exists x (vR^0x$ & $uRx)$.

We have explained that R^0 is the identity relation. So if vR^0x then $v=x$. But $\exists x (v=x$ & $uRx)$, is equivalent to uRv, and so the Euclidean condition is obtained:

$(wRv$ & $wRu) \Rightarrow uRv$.

In the case of axiom (4), $h=0$, $i=1$, $j=2$ and $k=0$. So the corresponding condition on frames is

$(w=v$ & $wR^2u) \Rightarrow \exists x (vRx$ & $u=x)$.

Resolving the identities this amounts to:

$vR^2u \Rightarrow vRu$.

By the definition of R^2, vR^2u iff $\exists x(vRx$ & $xRu)$, so this comes to:

$\exists x(vRx$ & $xRu) \Rightarrow vRu$,

which by predicate logic, is equivalent to transitivity.

vRx & $xRu \Rightarrow vRu$.

The reader may find it a pleasant exercise to see how the corresponding conditions fall out of hi-jk-Convergence when the values of the parameters h, i, j, and k are set by other axioms.

The Scott-Lemmon results provides a quick method for establishing results about the relationship between axioms and their corresponding frame conditions. Since they showed the adequacy of any logic that extends K with a selection of axioms of the form (G) with respect to models that satisfy the corresponding set of frame conditions, they provided "wholesale" adequacy proofs for the majority of systems in the modal family. Sahlqvist (1975) has discovered important generalizations of the Scott-Lemmon result covering a much wider range of axiom types.

The reader should be warned, however, that the neat correspondence between axioms and conditions on frames is atypical. There are condtions on frames that correspond to no axioms, and there are even conditions on frames for which no system is adequate.

Two Dimensional Semantics

Two dimensional semantics is a variant of possible world semantics that uses two (or more) kinds of parameters in truth evaluation, rather than possible worlds alone. For example, a logic of indexical expressions, such as 'I', 'here', 'now', and the like, needs to bring in the linguistic context (or context for short). Given a context $c = <s, p, t>$ where s is the speaker, p the place, and t the time of utterance, then 'I' refers to s, 'here' to p, and 'now' to t. So in the context c = <Jim Garson, Houston, 3:00 P.M. CST on 4/3/2014> 'I am here now' is T iff Jim Garson is in Houston, at 3:00 P.M. CST on 4/3/2014.

In possible worlds semantics, a sentence's truth-value depended on the world at which it is evaluated. However, indexicals bring in a second dimension – so we need to generalize again. Kaplan (1989) defines the character of a sentence B to be a function from the set of (linguistic) contexts to the content (or intension) of B, where the content, in turn, is simply the intension of B, that is a function from possible worlds to truth-values. Here, truth evaluation is doubly dependent - on both linguistic contexts and possible worlds.

One of Kaplan's most interesting observations is that some indexical sentences are contingent, but at the same time analytically true. An example is (1).

(1) I am here now.

Just from the meaning of the words, you can see that (1) must be true in any context $c = <s, p, t>$. After all, c counts as a linguistic context just in case s is a speaker who is at place p at time t. Therefore (1) is true at c, and that means that the pattern of truth-values (1) has along the context dimension must be all Ts (given the possible world is held fixed). This suggests that the context dimension is apt for tracking analytic knowledge obtained from the mastery of our language. On the other hand, the possible-worlds dimension keeps track of what is necessary. Holding the context fixed, there there are possible worlds where (1) is false. For example, when c =< Jim Garson, Houston, 3:00 P.M. CST on 4/3/2014>, (1) fails at c in a possible world where Jim Garson is in Boston at 3:00 P.M. CST on 4/3/2014. It follows that 'I am here now' is a contingent analytic truth. Therefore, two-dimensional semantics can handle situations where necessity and analyticity come apart.

Another example where bringing in two dimension is useful is in the logic for an open future (Thomason, 1984; Belnap, et al., 2001). Here one employs a temporal structure where many possible future histories extend from a given time. Consider (2).

(2) Joe will order a sea battle tomorrow.

If (2) is contingent, then there is a possible history where the battle occurs the day after the time of evaluation, and another one where it does not occur then. So to evaluate (2) you need to know two things: what is the time t of evaluation, and which of the histories h that run through t is the one to be considered. So a sentence in such a logic is evaluated at a pair $<t, h>$.

Another problem resolved by two-dimensional semantics is the interaction between 'now' and other temporal expressions like the future tense 'it will be the case that'. Then it is plausible to think that 'now' refers to the time of evaluation. So we would have the following truth condition:

> (Now) $v(NowB, t)$=T iff $v(B, t)$=T.

However this will not work for sentences like (3).

(3) At some point in the future, everyone now living will be unknown.

With F as the future tense operator, (3) might be translated:

> (3)' $F\forall x(NowLx \rightarrow Ux)$.

(The correct translation cannot be $\forall x(NowLx \rightarrow FUx)$, with F taking narrow scope, because (3) says there is a future time when all things now living are unknown together, not that each living thing will be unknown in some future time of its own). When the truth conditions for (3)' calculated, using (Now) and the truth condition (F) for F, it turns out that (3)' is true at time u iff there is a time t after u such that everything that is living at t (not u!) is unknown at t.

> (F) $v(FB, t)$=T iff for some time u later than t, $v(B, u)$=T.

To evaluate (3)' correctly so that it matches what we mean by (3), we must make sure that 'now' always refers back to the original time of utterance when 'now' lies in the scope of other temporal operators such as F. Therefore we need to keep track of which time is the time of utterance (u) as well as which time is the time of evaluation (t). So our indices take the form of a pair $<u, e>$, where u is the time of utterance, and e is the time of evaluation. Then the truth condition (Now) is revised to (2DNow).

> (2DNow) $v(NowB, <u, e>)$=T iff $v(B, <u, u>)$=T.

This has it that the NowB is true at a time u of utterance and time e of evaluation provided that B is true when u is taken to be the time of evaluation. When the truth conditions for F, $\forall$, and $\rightarrow$ are revised in the obvious way (just ignore the u in the pair), (3)' is true at $<u, e>$ provided that there is a time e' later than e such that everything that is living at u is unknown at e'. By carrying along a record of what u is during the truth calculation, we can always fix the value for 'now' to the original time of utterance, even when 'now' is deeply embedded in other temporal operators.

A similar phenomenon arises in modal logics with an actuality operator A (read 'it is actually the case that'). To properly evaluate (4) we need to keep track of which world is taken to be the actual (or real) world as well as which one is taken to the world of evaluation.

(4) It is possible that everyone actually living be unknown.

The idea of distinguishing different possible world dimensions in semantics has had useful applications in philosophy. For example, Chalmers (1996) has presented arguments from the conceivability of (say) zombies to dualist conclusions in the philosophy of mind. Chalmers (2006) has deployed two-dimensional semantics to help identify an a priori aspect of meaning that would support such conclusions.

The idea has also been deployed in the philosophy of language. Kripke (1980) famously argued that 'Water is H2O' is a posteriori but nevertheless a necessary truth, for given that water just is H2O, the there is no possible world where THAT stuff is (say) a basic element as the Greeks thought. On the other hand, there is a strong intuition that had the real world been somewhat different from what it is, the odorless liquid that falls from the sky as rain, fills our lakes and rivers, etc. might perfectly well have been an element. So in some sense it is conceivable that water is not H2O. Two dimensional semantics makes room for these intuitions by providing a separate dimension that tracks a conception of water that lays aside the chemical nature of what water actually is. Such a 'narrow content' account of the meaning of 'water' can explain how one may display semantical competence in the use of that term and still be ignorant about the chemistry of water (Chalmers, 2002).

Provability Logics

Modal logic has been useful in clarifying our understanding of central results concerning provability in the foundations of mathematics (Boolos, 1993). Provability logics are systems where the propositional variables p, q, r, etc. range over formulas of some mathematical system, for example Peano's system PA for arithmetic. (The system chosen for mathematics might vary, but assume it is PA for this discussion.) Gödel showed that arithmetic has strong expressive powers. Using code numbers for arithmetic sentences, he was able to demonstrate a correspondence between sentences of mathematics and facts about which sentences are and are not provable in PA. For example, he showed there there is a sentence C that is true just in case no contradiction is provable in PA and there is a sentence G (the famous Gödel sentence) that is true just in case it is not provable in PA.

In provability logics, $\Box p$ is interpreted as a formula (of arithmetic) that expresses that what p denotes is provable in PA. Using this notation, sentences of provability logic express facts about provability. Suppose that $\bot$ is a constant of provability logic denoting a contradiction. Then $\sim\Box\bot$ says that PA is consistent and $\Box A \rightarrow A$ says that PA is sound in the sense that when it proves A, A is indeed true. Furthermore, the box may be iterated. So, for example, $\Box\sim\Box\bot$ makes the dubious claim that PA is able to prove its own consistency, and $\sim\Box\bot \rightarrow \sim\Box\sim\Box\bot$ asserts (correctly as Gödel proved) that if PA is consistent then PA is unable to prove its own consistency.

Although provability logics form a family of related systems, the system GL is by far the best known. It results from adding the following axiom to K:

(GL) $\Box(\Box A \rightarrow A) \rightarrow \Box A$

The axiom (4): □A→□□A is provable in GL, so GL is actually a strengthening of K4. However, axioms such as (M): □A→A, and even the weaker (D): □A→◇A are not available (nor desirable) in GL. In provability logic, provability is not to be treated as a brand of necessity. The reason is that when p is provable in an arbitrary system S for mathematics, it does not follow that p is true, since S may be unsound. Furthermore, if p is provable in S (□p) it need not even follow that ~p lacks a proof (~□~p = ◇p). S might be inconsistent and so prove both p and ~p.

Axiom (GL) captures the content of Löb's Theorem, an important result in the foundations of arithmetic. □A→A says that PA is sound for A, i.e. that if A were proven, A would be true. (Such a claim might not be secure for an arbitrarily selected system S, since A might be provable in S and false.) (GL) claims that if PA manages to prove the sentence that claims soundness for a given sentence A, then A is already provable in PA. Löb's Theorem reports a kind of modesty on PA's part. PA never insists (proves) that a proof of A entails A's truth, unless it already has a proof of A to back up that claim.

It has been shown that GL is adequate for provability in the following sense. Let a sentence of GL be always provable exactly when the sentence of arithmetic it denotes is provable no matter how its variables are assigned values to sentences of PA. Then the provable sentences of GL are exactly the sentences that are always provable. This adequacy result has been extremely useful, since general questions concerning provability in PA can be transformed into easier questions about what can be demonstrated in GL.

GL can also be outfitted with a possible world semantics for which it is sound and complete. A corresponding condition on frames for GL-validity is that the frame be transitive, finite and irreflexive.

Advanced Modal Logic

The applications of modal logic to mathematics and computer science have become increasingly important. Provability logic is only one example of this trend. The term "advanced modal logic" refers to a tradition in modal logic research that is particularly well represented in departments of mathematics and computer science. This tradition has been woven into the history of modal logic right from its beginnings (Goldblatt, 2006). Research into relationships with topology and algebras represents some of the very first technical work on modal logic. However the term 'advanced modal logic' generally refers to a second wave of work done since the mid 1970s. Some examples of the many interesting topics dealt with include results on decidability (whether it is possible to compute whether a formula of a given modal logic is a theorem) and complexity (the costs in time and memory needed to compute such facts about modal logics).

Bisimulation

Bisimulation provides a good example of the fruitful interactions that have been developed between modal logic and computer science. In computer science, labeled transition systems (LTSs) are commonly used to represent possible computation pathways during execution of a program. LTSs are generalizations of Kripke frames, consisting of a set W of states, and a collection of i-accessibility relations R_i, one for each computer process i. Intuitively, wR_iw' holds exactly when w' is a state that results from applying the process i to state w.

The language of poly-modal or dynamic logic introduces a collection of modal operators $\square_i$, one for each program i (Harel, 1984). Then $\square_i A$ states that sentence A holds in every result of applying i. So ideas like the correctness and successful termination of programs can be expressed in this language. Models for such a language are like Kripke models save that LTSs are used in place of frames. A bisimulation is a counterpart relation between states of two such models such that exactly the same propositional variables are true in counterpart states, and whenever world v is i-accessible from one of two counterpart states, then the other counterpart bears the i-accessibility relation to some counterpart of v. In short, the i-accessibility structure one can "see" from a given state mimics what one sees from a counterpart. Bisimulation is a weaker notion than isomorphism (a bisimulation relation need not be 1-1), but it is sufficient to guarantee equivalence in processing.

In the 1970s, a version of bisimulation had already been developed by modal logicians to help better understand the relationship between modal logic axioms and their corresponding conditions on Kripke frames. Kripke's semantics provides a basis for translating modal axioms into sentences of a second-order language where quantification is allowed over one-place predicate letters P. Replace metavariables A with open sentences Px, translate $\square$Px to $\forall y(Rxy \rightarrow Py)$, and close free variables x and predicate letters P with universal quantifiers. For example, the predicate logic translation of the axiom schema $\square A \rightarrow A$ comes to $\forall P \forall x[\forall y(Rxy \rightarrow Py) \rightarrow Px]$. Given this translation, one may instantiate the variable P to an arbitrary one-place predicate, for example to the predicate Rx whose extension is the set of all worlds w such that Rxw for a given value of x. Then one obtains $\forall x[\forall y(Rxy \rightarrow Rxy) \rightarrow Rxx]$, which reduces to $\forall xRxx$, since $\forall y(Rxy \rightarrow Rxy)$ is a tautology. This illuminates the correspondence between $\square A \rightarrow A$ and reflexivity of frames ($\forall xRxx$). Similar results hold for many other axioms and frame conditions. The "collapse" of second-order axiom conditions to first order frame conditions is very helpful in obtaining completeness results for modal logics. For example, this is the core idea behind the elegant results of Sahlqvist (1975).

But when does the second-order translation of an axiom reduce to a first-order condition on R in this way? In the 1970s, van Benthem showed that this happens iff the translation's holding in a model entails its holding in any bisimular model, where two models are bisimular iff there is a bisimulation between them in the special case where there is a single accessibility relation. That result generalizes easily to the poly-modal case. This suggests that poly-modal logic lies at exactly the right level of abstraction to describe, and reason about, computation and other processes. (After all, what really matters there is the preservation of truth values of formulas in models rather than the finer details of the frame structures.) Furthermore the implicit translation of those logics into well-understood fragments of predicate logic provides a wealth of information of interest to computer scientists. As a result, a fruitful area of research in computer science has developed with bisimulation as its core idea.

Quantifiers in Modal Logic

It would seem to be a simple matter to outfit a modal logic with the quantifiers $\forall$ (all) and $\exists$ (some). One would simply add the standard (or classical) rules for quantifiers to the principles of whichever propositional modal logic one chooses. However, adding quantifiers to modal logic involves a number of difficulties. Some of these are philosophical. For example, Quine (1953) has famously argued that quantifying into modal contexts is simply incoherent, a view that has spawned a gigantic literature. Quine's complaints do not carry the weight they once did. Nevertheless, the view that there is something wrong with "quantifying in" is still widely held. A second kind of complication is technical. There is a wide variety in the choices one can make in the semantics for quantified modal logic, and the proof that a system of rules is correct for a given choice can be difficult. The work of Corsi (2002) and Garson (2005) goes some way towards bringing unity to this terrain, but the situation still remains challenging.

Another complication is that some logicians believe that modality requires abandoning classical quantifier rules in favor of the weaker rules of free logic (Garson 2001). The main points of disagreement concerning the quantifier rules can be traced back to decisions about how to handle the domain of quantification. The simplest alternative, the fixed-domain (sometimes called the possibilist) approach, assumes a single domain of quantification that contains all the possible objects. On the other hand, the world-relative (or actualist) interpretation, assumes that the domain of quantification changes from world to world, and contains only the objects that actually exist in a given world.

The fixed-domain approach requires no major adjustments to the classical machinery for the quantifiers. Modal logics that are adequate for fixed domain semantics can usually be axiomatized by adding principles of a propositional modal logic to classical quantifier rules together with the Barcan Formula (BF) (Barcan 1946).

(BF) $\forall x \Box A \rightarrow \Box \forall x A$.

The fixed-domain interpretation has advantages of simplicity and familiarity, but it does not provide a direct account of the semantics of certain quantifier expressions of natural language. We do not think that 'Some man exists who signed the Declaration of Independence' is true, at least not if we read 'exists' in the present tense. Nevertheless, this sentence was true in 1777, which shows that the domain for the natural language expression 'some man exists who' changes to reflect which men exist at different times. A related problem is that on the fixed-domain interpretation, the sentence $\forall y \Box \exists x(x=y)$ is valid. Assuming that $\exists x(x=y)$ is read: y exists, $\forall y \Box \exists x(x=y)$ says that everything exists necessarily. However, it seems a fundamental feature of common ideas about modality that the existence of many things is contingent, and that different objects exist in different possible worlds.

The defender of the fixed-domain interpretation may respond to these objections by insisting that on his (her) reading of the quantifiers, the domain of quantification contains all possible objects, not just the objects that happen to exist at a given world. So the theorem $\forall y \Box \exists x(x=y)$ makes the innocuous claim that every possible object is necessarily found in the domain of all possible objects. Furthermore, those quantifier expressions of natural language whose domain is world (or time) dependent can be expressed using the fixed-domain quantifier $\exists x$ and a predicate letter E with the reading 'actually exists'. For example, instead of translating 'Some Man exists who Signed the Declaration of Independence' by

$\exists x(Mx \& Sx)$,

the defender of fixed domains may write:

$\exists x(Ex \& Mx \& Sx)$,

thus ensuring the translation is counted false at the present time. Cresswell (1991) makes the interesting observation that world-relative quantification has limited expressive power relative to fixed-domain quantification. World-relative quantification can be defined with fixed domain quantifiers and E, but there is no way to fully express fixed-domain quantifiers with world-relative ones. Although this argues in favor of the classical approach to quantified modal logic, the translation tactic also amounts to something of a concession in favor of free logic, for the world-relative quantifiers so defined obey exactly the free logic rules.

A problem with the translation strategy used by defenders of fixed domain quantification is that rendering the English into logic is less direct, since E must be added to all translations of all sentences whose quantifier expressions have domains that are context dependent. A more serious objection to fixed-domain quantification is that it strips the quantifier of a role which Quine recommended for it, namely to record robust ontological commitment. On this view, the domain of $\exists x$ must contain only entities that are ontologically respectable, and possible objects are too abstract to qualify. Actualists of this stripe will want to develop the logic of a quantifier $\exists x$ which reflects commitment to what is actual in a given world rather than to what is merely possible.

However, recent work on actualism tends to undermine this objection. For example, Linsky and Zalta (1994) argue that the fixed-domain quantifier can be given an interpretation that is perfectly acceptable to actualists. Actualists who employ possible worlds semantics routinely quantify over possible worlds in their semantical theory of language. So it would seem that possible worlds are actual by these actualist's lights. By cleverly outfitting the domain with abstract entities no more objectionable than the ones actualists accept, Linsky and Zalta show that the Barcan Formula and classical principles can be vindicated. Note however, that actualists may respond that they need not be committed to the actuality of possible worlds so long as it is understood that quantifiers used in their theory of language lack strong ontological import. In any case, it is open to actualists (and non actualists as well) to investigate the logic of quantifiers with more robust domains, for example domains excluding possible worlds and other such abstract entities, and containing only the spatio-temporal particulars found in a given world. For quantifiers of this kind, a world-relative domains are appropriate.

Such considerations motivate interest in systems that acknowledge the context dependence of quantification by introducing world-relative domains. Here each possible world has its own domain of quantification (the set of objects that actually exist in that world), and the domains vary from one world to the next. When this decision is made, a difficulty arises for classical quantification theory. Notice that the sentence $\exists x(x=t)$ is a theorem of classical logic, and so $\Box\exists x(x=t)$ is a theorem of K by the Necessitation Rule. Let the term t stand for Saul Kripke. Then this theorem says that it is necessary that Saul Kripke exists, so that he is in the domain of every possible world. The whole motivation for the world-relative approach was to reflect the idea that objects in one world may fail to exist in another. If standard quantifier rulers are used, however, every term t must refer to something that exists in all the possible worlds. This seems incompatible with our ordinary practice of using terms to refer to things that only exist contingently.

One response to this difficulty is simply to eliminate terms. Kripke (1963) gives an example of a system that uses the world-relative interpretation and preserves the classical rules. However, the costs are severe. First, his language is artificially impoverished, and second, the rules for the propositional modal logic must be weakened.

Presuming that we would like a language that includes terms, and that classical rules are to be added to standard systems of propositional modal logic, a new problem arises. In such a system, it is possible to prove (CBF), the converse of the Barcan Formula.

(CBF) □∀xA→∀x□A.

This fact has serious consequences for the system's semantics. It is not difficult to show that every world-relative model of (CBF) must meet condition (ND) (for 'nested domains').

(ND) If wRv then the domain of w is a subset of the domain of v.

However (ND) conflicts with the point of introducing world-relative domains. The whole idea was that existence of objects is contingent so that there are accessible possible worlds where one of the things in our world fails to exist.

A straightforward solution to these problems is to abandon classical rules for the quantifiers and to adopt rules for free logic (FL) instead. The rules of FL are the same as the classical rules, except that inferences from ∀xRx (everything is real) to Rp (Pegasus is real) are blocked. This is done by introducing a predicate 'E' (for 'actually exists') and modifying the rule of universal instantiation. From ∀xRx one is allowed to obtain Rp only if one also has obtained Ep. Assuming that the universal quantifier ∀x is primitive, and the existential quantifier ∃x is defined by $∃xA =_{df} ~∀x~A$, then FL may be constructed by adding the following two principles to the rules of propositional logic

Universal Generalization. If B→(Ey→A(y)) is a theorem, so is B→∀xA(x).

Universal Instantiation. ∀xA(x)→(En→A(n))

(Here it is assumed that A(x) is any well-formed formula of predicate logic, and that A(y) and A(n) result from replacing y and n properly for each occurrence of x in A(x).) Note that the instantiation axiom is restricted by mention of En in the antecedent. The rule of Universal Generalization is modified in the same way. In FL, proofs of formulas like ∃x□(x=t), ∀y□ ∃x(x=y), (CBF), and (BF), which seem incompatible with the world-relative interpretation, are blocked.

One philosophical objection to FL is that E appears to be an existence predicate, and many would argue that existence is not a legitimate property like being green or weighing more than four pounds. So philosophers who reject the idea that existence is a predicate may object to FL. However in most (but not all) quantified modal logics that include identity (=) these worries may be skirted by defining E as follows.

$Et =_{df} ∃x(x=t)$.

The most general way to formulate quantified modal logic is to create FS by adding the rules of FL to a given propositional modal logic S. In situations where classical quantification is desired, one may simply add Et as an axiom to FS, so that the classical principles become derivable rules. Adequacy results for such systems can be obtained for most choices of the modal logic S, but there are exceptions.

A final complication in the semantics for quantified modal logic is worth mentioning. It arises when non-rigid expressions such as 'the inventor of bifocals' are introduced to the language. A term is non-rigid when it picks out different objects in different possible worlds. The semantical value of such a term can be given by what Carnap (1947) called an individual concept, a function that picks out the denotation of the term for each possible world. One approach to dealing with non-rigid terms is to employ Russell's theory of descriptions. However, in a language that treats non rigid expressions as genuine terms, it turns out that neither the classical nor the free logic rules for the quantifiers are acceptable. (The problem can not be resolved by weakening the rule of substitution for identity.) A solution to this problem is to employ a more general treatment of the quantifiers, where the domain of quantification contains individual concepts rather than objects. This more general interpretation provides a better match between the treatment of terms and the treatment of quantifiers and results in systems that are adequate for classical or free logic rules (depending on whether the fixed domains or world-relative domains are chosen). It also provides a language with strong and much needed expressive powers.

References

- Hunter, Geoffrey (1971). Metalogic: An Introduction to the Metatheory of Standard First-Order Logic. University of California Pres. ISBN 0-520-02356-0

- Discrete-mathematics, discrete-mathematics-propositional-logic: tutorialspoint.com, Retrieved 30 June 2018

- Anellis, Irving H. (2012). "Peirce's Truth-functional Analysis and the Origin of the Truth Table". History and Philosophy of Logic. 33: 87–97. doi:10.1080/01445340.2011.621702

- Logic-infinitary: plato.stanford.edu, Retrieved 23 July 2018

- Toida, Shunichi (2 August 2009). "Proof of Implications". CS381 Discrete Structures/Discrete Mathematics Web Course Material. Department Of Computer Science, Old Dominion University. Retrieved 10 March 2010

- What-is-boolean-logic-definition-diagram-examples: study.com, Retrieved 27 June 2018

- Bennett, David (1980). "Junctions". Notre Dame Journal of Formal Logic. XXI (1): 111–118. doi:10.1305/ndj-fl/1093882943

- Logic-non-classical, encyclopedias-almanacs-transcripts-and-maps: encyclopedia.com, Retrieved 18 May 2018

- Collinson, Matthew; Monahan, Brian; Pym, David (2012). A Discipline of Mathematical Systems Modelling. London: College Publications. ISBN 978-1-904987-50-5

- Modal-logic: britannica.com, Retrieved 28 May 2018

5

Set Theory

Set theory is a subfield of mathematical logic, which is concerned with the study of sets. All the diverse aspects of set theory have been carefully analyzed in this chapter, such as axioms of set theory, set theory of the continuum, countable and uncountable sets, Zermelo-Fraenkel set theory and Cantor's theorem.

Set theory is the mathematical theory of well-determined collections, called sets, of objects that are called members, or elements, of the set. Pure set theory deals exclusively with sets, so the only sets under consideration are those whose members are also sets. The theory of the hereditarily-finite sets, namely those finite sets whose elements are also finite sets, the elements of which are also finite, and so on, is formally equivalent to arithmetic. So, the essence of set theory is the study of infinite sets, and therefore it can be defined as the mathematical theory of the actual—as opposed to potential—infinite.

The notion of set is so simple that it is usually introduced informally, and regarded as self-evident. In set theory, however, as is usual in mathematics, sets are given axiomatically, so their existence and basic properties are postulated by the appropriate formal axioms. The axioms of set theory imply the existence of a set-theoretic universe so rich that all mathematical objects can be construed as sets. Also, the formal language of pure set theory allows one to formalize all mathematical notions and arguments. Thus, set theory has become the standard foundation for mathematics, as every mathematical object can be viewed as a set, and every theorem of mathematics can be logically deduced in the Predicate Calculus from the axioms of set theory.

A Set is a unordered collection of objects, known as elements or members of the set.

An element 'a' belong to a set A can be written as 'a $\in$ A', 'a $\notin$ A' denotes that a is not an element of the set A.

Representation of a Set

A set can be represented by various methods. 3 common methods used for representing set:

1. Statement form.

2. Roaster form or tabular form method.

3. Set Builder method.

Statement Form

In this representation, well defined description of the elements of the set is given. Below are some examples of same.

1. The set of all even number less than 10.

2. The set of number less than 10 and more than 1.

Roster Form

In this representation, elements are listed within the pair of brackets {} and are separated by commas. Below are two examples.

1. Let N is the set of natural numbers less than 5.

   N = { 1 , 2 , 3, 4 }.

2. The set of all vowels in english alphabet.

   V = { a , e , i , o , u }.

Set Builder Form

In set builder set is describe by a property that its member must satisfy.

1. {x : x is even number divisible by 6 and less than 100}.

2. {x : x is natural number less than 10}.

Equal Sets

Two sets are said to be equal if both have same elements. For example A = {1, 3, 9, 7} and B = {3, 1, 7, 9} are equal sets.

Note: Order of elements of a set doesn't matter.

Universal Set

It's a set that contains everything. Well, not *exactly* everything. Everything that is relevant to our question.

1 5 3400 -3 -5 1000 15	Then our sets included integers. The universal set for that is all the integers. In fact, when doing Number Theory, this is almost always what the universal set is, as Number Theory is simply the study of integers.
1 -3.6 -5 1000 3.33333 15 0.001	But in Calculus (also known as real analysis), the universal set is almost always the real numbers.
1 i $-3+i$ -5 $2i$ 3400 15 $10+3i$	And in complex analysis, you guessed it, the universal set is the complex numbers.

Equality

Two sets are equal if they have precisely the same members. Now, at first glance they may not seem equal, so we may have to examine them closely.

Example: Are A and B equal where:

- A is the set whose members are the first four positive whole numbers

- B = {4, 2, 1, 3}

Let's check. They both contain 1. They both contain 2. And 3, And 4. And we have checked every element of both sets, so: Yes, they are equal.

And the equals sign (=) is used to show equality, so we write:

A = B

Example: Are these sets equal?

- A is {1, 2, 3}

- B is {3, 1, 2}

Yes, they are equal.

They both contain exactly the members 1, 2 and 3.

It doesn't matter *where* each member appears, so long as it is there.

Subsets

When we define a set, if we take pieces of that set, we can form what is called a subset.

Example: the set {1, 2, 3, 4, 5}

A subset of this is {1, 2, 3}. Another subset is {3, 4} or even another is {1}, etc.

But {1, 6} is not a subset, since it has an element (6) which is not in the parent set.

In general:

A is a subset of B if and only if every element of A is in B.

So let's use this definition in some examples.

Example: Is A a subset of B, where A = {1, 3, 4} and B = {1, 4, 3, 2}?

1 is in A, and 1 is in B as well. So far so good.

3 is in A and 3 is also in B.

4 is in A, and 4 is in B.

That's all the elements of A, and every single one is in B, so we're done.

> Yes, A is a subset of B

Note that 2 is in B, but 2 is not in A. But remember, that doesn't matter, we only look at the elements in A.

Let's try a harder example.

Example: Let A be all multiples of 4 and B be all multiples of 2.

Is A a subset of B? And is B a subset of A?

Well, we can't check every element in these sets, because they have an infinite number of elements. So we need to get an idea of what the elements look like in each, and then compare them.

The sets are:

- A = {..., -8, -4, 0, 4, 8, ...}
- B = {..., -8, -6, -4, -2, 0, 2, 4, 6, 8, ...}

By pairing off members of the two sets, we can see that every member of A is also a member of B, but not every member of B is a member of A:

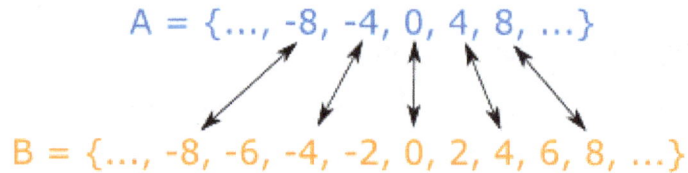

$$A = \{..., -8, -4, 0, 4, 8, ...\}$$

$$B = \{..., -8, -6, -4, -2, 0, 2, 4, 6, 8, ...\}$$

So:

> A is a subset of B, but B is not a subset of A

Proper Subsets

If we look at the defintion of subsets and let our mind wander a bit, we come to a weird conclusion.

Let A be a set. Is every element in A an element in A?

> Well, umm, *yes of course*, right?

So doesn't that mean that *A is a subset of A*?

This doesn't seem very *proper*, does it? We want our subsets to be *proper*. So we introduce (what else but) proper subsets.

A is a proper subset of B if and only if every element in A is also in B, and there exists at least one element in B that is not in A.

This little piece at the end is only there to make sure that A is not a proper subset of itself. Otherwise, a proper subset is exactly the same as a normal subset.

Example

{1, 2, 3} is a subset of {1, 2, 3}, but is not a proper subset of {1, 2, 3}.

Example

{1, 2, 3} is a proper subset of {1, 2, 3, 4} because the element 4 is not in the first set.

Notice that if A is a proper subset of B, then it is also a subset of B.

Empty (Null) Set

This is probably the weirdest thing about sets.

As an example, think of the set of piano keys on a guitar.

"But wait!" you say, "There are no piano keys on a guitar!"

And right you are. It is a set with no elements.

This is known as the Empty Set (or Null Set).There aren't any elements in it. Not one. Zero.

It is represented by ∅

Or by {} (a set with no elements)

Some other examples of the empty set are the set of countries south of the south pole.

So what's so weird about the empty set? Well, that part comes next.

Empty Set and Subsets

So let's go back to our definition of subsets. We have a set A. We won't define it any more than that, it could be any set. *Is the empty set a subset of A?*

Going back to our definition of subsets, if every element in the empty set is also in A, then the empty set is a subset of A. But what if we have no elements?

It takes an introduction to logic to understand this, but this statement is one that is "vacuously" or "trivially" true.

A good way to think about it is: *we can't find any elements in the empty set that aren't in A*, so it must be that all elements in the empty set are in A.

So the answer to the posed question is a resounding yes.

The empty set is a subset of every set, including the empty set itself.

Size of a Set

Size of a set can be finite or infinite.

For example

```
Finite set: Set of natural numbers less than 100.

Infinite set: Set of real numbers.
```

Size of the set S is known as Cardinality number, denoted as $|S|$.

Example: Let A be a set of odd positive integers less than 10.

Solution : A = {1, 3, 5, 7, 9}, Cardinality of the set is 5, i.e.,$|A| = 5$.

Note: Cardinality of null set is 0.

Power Sets

Power set is the set all possible subset of the set S. Denoted by P(S).

Example : What is the power set of {0,1,2}?

Solution: All possible subsets

{∅}, {0}, {1}, {2}, {0,1}, {0,2}, {1,2}, {0,1,2}.

Note: Empty set and set itself is also member of this set of subsets.

Cardinality of power set is

$$2^n$$

where n is number of element in a set.

Cartesian Products

Let A and B be two sets. Cartesian product of A and B is denoted by A × B, is the set of all ordered pairs (a,b), where a belong to A and B belong to B.

A × B = {(a, b) | a ∈ A ∧ b ∈ B}.

Example 1. What is Cartesian product of A = {1,2} and B = {p,q,r}.

Solution : A × B ={(1,p), (1,q), (1,r), (2,p), (2,q), (2,r) };

Cardinality of A × B is N*M, where N is the Cardinality of A and M is the cardinality of B.

Note : A × B is not same as B × A.

Areas of Study

Set theory is a major area of research in mathematics, with many interrelated subfields.

Combinatorial Set Theory

Combinatorial set theory concerns extensions of finite combinatorics to infinite sets. This includes the study of cardinal arithmetic and the study of extensions of Ramsey's theorem such as the Erdős–Rado theorem.

Descriptive Set Theory

Descriptive set theory is the study of subsets of the real line and, more generally, subsets of Polish spaces. It begins with the study of pointclasses in the Borel hierarchy and extends to the study of more complex hierarchies such as the projective hierarchy and the Wadge hierarchy. Many properties of Borel sets can be established in ZFC, but proving these properties hold for more complicated sets requires additional axioms related to determinacy and large cardinals.

The field of effective descriptive set theory is between set theory and recursion theory. It includes the study of lightface pointclasses, and is closely related to hyperarithmetical theory. In many cases, results of classical descriptive set theory have effective versions; in some cases, new results are obtained by proving the effective version first and then extending ("relativizing") it to make it more broadly applicable.

A recent area of research concerns Borel equivalence relations and more complicated definable equivalence relations. This has important applications to the study of invariants in many fields of mathematics.

Fuzzy Set Theory

In set theory as Cantor defined and Zermelo and Fraenkel axiomatized, an object is either a member of a set or not. In fuzzy set theory this condition was relaxed by Lotfi A. Zadeh so an object has a *degree of membership* in a set, a number between 0 and 1. For example, the degree of membership of a person in the set of "tall people" is more flexible than a simple yes or no answer and can be a real number such as 0.75.

Inner Model Theory

An inner model of Zermelo–Fraenkel set theory (ZF) is a transitive class that includes all the ordinals and satisfies all the axioms of ZF. The canonical example is the constructible universe *L* developed by Gödel. One reason that the study of inner models is of interest is that it can be used to prove consistency results. For example, it can be shown that regardless of whether a model *V* of ZF satisfies the continuum hypothesis or the axiom of choice, the inner model *L* constructed inside the original model will satisfy both the generalized continuum hypothesis and the axiom of choice. Thus the assumption that ZF is consistent (has at least one model) implies that ZF together with these two principles is consistent.

The study of inner models is common in the study of determinacy and large cardinals, especially when considering axioms such as the axiom of determinacy that contradict the axiom of choice. Even if a fixed model of set theory satisfies the axiom of choice, it is possible for an inner model to fail to satisfy the axiom of choice. For example, the existence of sufficiently large cardinals implies that there is an inner model satisfying the axiom of determinacy (and thus not satisfying the axiom of choice).

Large Cardinals

A large cardinal is a cardinal number with an extra property. Many such properties are studied, including inaccessible cardinals, measurable cardinals, and many more. These properties typically imply the cardinal number must be very large, with the existence of a cardinal with the specified property unprovable in Zermelo-Fraenkel set theory.

Determinacy

Determinacy refers to the fact that, under appropriate assumptions, certain two-player games of perfect information are determined from the start in the sense that one player must have a winning strategy. The existence of these strategies has important consequences in descriptive set theory, as the assumption that a broader class of games is determined often implies that a broader class of sets will have a topological property. The axiom of determinacy (AD) is an important object of study; although incompatible with the axiom of choice, AD implies that all subsets of the real line are well behaved (in particular, measurable and with the perfect set property). AD can be used to prove that the Wadge degrees have an elegant structure.

Forcing

Paul Cohen invented the method of forcing while searching for a model of ZFC in which the continuum hypothesis fails, or a model of ZF in which the axiom of choice fails. Forcing adjoins to some given model of set theory additional sets in order to create a larger model with properties determined (i.e. "forced") by the construction and the original model. For example, Cohen's construction adjoins additional subsets of the natural numbers without changing any of the cardinal numbers of the original model. Forcing is also one of two methods for proving relative consistency by finitistic methods, the other method being Boolean-valued models.

Cardinal Invariants

A cardinal invariant is a property of the real line measured by a cardinal number. For example, a well-studied invariant is the smallest cardinality of a collection of meagre sets of reals whose union is the entire real line. These are invariants in the sense that any two isomorphic models of set theory must give the same cardinal for each invariant. Many cardinal invariants have been studied, and the relationships between them are often complex and related to axioms of set theory.

Set-theoretic Topology

Set-theoretic topology studies questions of general topology that are set-theoretic in nature or that require advanced methods of set theory for their solution. Many of these theorems are independent of ZFC, requiring stronger axioms for their proof. A famous problem is the normal Moore space question, a question in general topology that was the subject of intense research. The answer to the normal Moore space question was eventually proved to be independent of ZFC.

Operation on Set

When two or more sets combine together to form one set under the given conditions, then operations on sets are carried out.

Four basic operations on set

The four basic operations are:

1.   Union of Sets

2.   Intersection of sets

3.   Complement of the Set

4.   Cartesian Product of sets

Union of Sets

Union of two given sets is the smallest set which contains all the elements of both the sets.

To find the union of two given sets A and B is a set which consists of all the elements of A and all the elements of B such that no element is repeated.

The symbol for denoting union of sets is '∪'.

For example

    Let set A = {2, 4, 5, 6}

    and set B = {4, 6, 7, 8}

Taking every element of both the sets A and B, without repeating any element, we get a new set = {2, 4, 5, 6, 7, 8}

This new set contains all the elements of set A and all the elements of set B with no repetition of elements and is named as union of set A and B.

The symbol used for the union of two sets is '∪'.

Therefore, symbolically, we write union of the two sets A and B is A ∪ B which means A union B.

Therefore, $A \cup B = \{x : x \in A \text{ or } x \in B\}$.

Intersection of Sets

Intersection of two given sets is the largest set which contains all the elements that are common to both the sets.

To find the intersection of two given sets A and B is a set which consists of all the elements which are common to both A and B.

The symbol for denoting intersection of sets is '∩'.

For example:

Let set A = {2, 3, 4, 5, 6}

and set B = {3, 5, 7, 9}

In this two sets, the elements 3 and 5 are common. The set containing these common elements i.e., {3, 5} is the intersection of set A and B.

The symbol used for the intersection of two sets is '∩'.

Therefore, symbolically, we write intersection of the two sets A and B is A ∩ B which means A intersection B.

The intersection of two sets A and B is represented as A ∩ B = {x : x ∈ A and x ∈ B} .

How to find the difference of two sets?

If A and B are two sets, then their difference is given by A - B or B - A.

- If A = {2, 3, 4} and B = {4, 5, 6}

  A - B means elements of A which are not the elements of B.

  i.e., in the above example A - B = {2, 3}

  In general, B - A = {x : x ∈ B, and x ∉ A}

- If A and B are disjoint sets, then A − B = A and B − A = B

Complement of a Set

In complement of a set if ξ be the universal set and A a subset of ξ, then the complement of A is the set of all elements of ξ which are not the elements of A.

Symbolically, we denote the complement of A with respect to ξ as A'.

For example:

If ξ = {1, 2, 3, 4, 5, 6, 7}

A = {1, 3, 7} find A'.

Solution:

We observe that 2, 4, 5, 6 are the only elements of ξ which do not belong to A.

Therefore, A' = {2, 4, 5, 6}

Note: The complement of a universal set is an empty set.

The complement of an empty set is a universal set.

The set and its complement are disjoint sets.

For example:

1. Let the set of natural numbers be the universal set and A is a set of even natural numbers,

then A' {x: x is a set of odd natural numbers}

2. Let ξ = The set of letters in the English alphabet.

A = The set of consonants in the English alphabet

then A' = The set of vowels in the English alphabet.

3. Show that;

(a) The complement of a universal set is an empty set.

Let ξ denote the universal set, then

ξ' = The set of those elements which are not in ξ.

= empty set = ϕ

Therefore, ξ = ϕ so the complement of a universal set is an empty set.

(b) A set and its complement are disjoint sets.

Let A be any set then A' = set of those elements of ξ which are not in A'.

Let x ∉ A, then x is an element of ξ not contained in A'

So x ∉ A'

Therefore, A and A' are disjoint sets.

Therefore, Set and its complement are disjoint sets.

Similarly, in complement of a set when U be the universal set and A is a subset of U. Then the complement of A is the set all elements of U which are not the elements of A.

Symbolically, we write A' to denote the complement of A with respect to U.

Thus, A' = {x : x ∈ U and x ∉ A}

Obviously A' = {U - A}

For example: Let U = {2, 4, 6, 8, 10, 12, 14, 16}

A = {6, 10, 4, 16}

A' = {2, 8, 12, 14}

We observe that 2, 8, 12, 14 are the only elements of U which do not belong to A.

Some properties of complement sets

- A ∪ A' = A' ∪ A = ∪ (Complement law)
- (A ∩ B') = φ (Complement law)
- (A ∪ B) = A' ∩ B' (De Morgan's law)
- (A ∩ B)' = A' ∪ B' (De Morgan's law)
- (A')' = A (Law of complementation)
- φ' = ∪ (Law of empty set
- ∪' = φ and universal set)

Axioms of Set Theory

In contrast to naive set theory, the attitude adopted in an axiomatic development of set theory is that it is not necessary to know what the "things" are that are called "sets" or what the relation of membership means. Of sole concern are the properties assumed about sets and the membership relation. Thus, in an axiomatic theory of sets, set and the membership relation ∈ are undefined terms. The assumptions adopted about these notions are called the axioms of the theory. Axiomatic set theorems are the axioms together with statements that can be deduced from the axioms using the rules of inference provided by a system of logic. Criteria for the choice of axioms include:

(1) Consistency—it should be impossible to derive as theorems both a statement and its negation;

(2) Plausibility—axioms should be in accord with intuitive beliefs about sets; and

(3) Richness—desirable results of Cantorian set theory can be derived as theorems.

Russell's Paradox

Russell's paradox, statement in set theory, devised by the English mathematician-philosopher Bertrand Russell, that demonstrated a flaw in earlier efforts to axiomatize the subject.

Russell found the paradox in 1901 and communicated it in a letter to the German mathematician-logician Gottlob Frege in 1902. Russell's letter demonstrated an inconsistency in Frege's axiomatic system of set theory by deriving a paradox within it. (The German mathematician Ernst Zermelo had found the same paradox independently; since it could not be produced in his own axiomatic system of set theory, he did not publish the paradox.)

Frege had constructed a logical system employing an unrestricted comprehension principle. The comprehension principle is the statement that, given any condition expressible by a formula $\phi(x)$, it is possible to form the set of all sets x meeting that condition, denoted $\{x \mid \phi(x)\}$. For example, the set of all sets—the universal set—would be $\{x \mid x = x\}$.

It was noticed in the early days of set theory, however, that a completely unrestricted comprehension principle led to serious difficulties. In particular, Russell observed that it allowed the formation of $\{x \mid x \notin x\}$, the set of all non-self-membered sets, by taking $\phi(x)$ to be the formula $x \notin x$. Is this set—call it R—a member of itself? If it is a member of itself, then it must meet the condition of its not being a member of itself. But if it is not a member of itself, then it precisely meets the condition of being a member of itself. This impossible situation is called Russell's paradox.

The significance of Russell's paradox is that it demonstrates in a simple and convincing way that one cannot both hold that there is meaningful totality of all sets and also allow an unfettered comprehension principle to construct sets that must then belong to that totality. (Russell spoke of this situation as a "vicious circle.")

Set theory avoids this paradox by imposing restrictions on the comprehension principle. The standard Zermelo-Fraenkel axiomatization.

Zermelo-Fraenkel axioms

(1) *Axiom of extension.* If A and B are sets and if, for all x, x ∈ A if and only if x ∈ B, then A = B.

(2) *Axiom of the empty set.* There exists a set A such that, for all x, it is false that x ∈ A.

(3) *Axiom schema of separation.* If A is a set, there exists a set B such that, for all x, x ∈ B if and only if x ∈ A and S(x). Here, S(x) is any condition on x in which B is not free (it must be bound by a quantifier such as "all" or "some").

(4) *Axiom of pairing.* If A and B are sets, there exists a set (symbolized {A, B} and called the unordered pair of A and B) having A and B as its sole members.

(5) *Axiom of union.* If C is a set, there exists a set A such that x ∈ A if and only if x ∈ B for some member B of C.

(6) *Axiom of power set.* If A is a set, there exists a set B, called its power set, such that x ∈ B if and only if x ⊆ A.

(7) *Axiom of infinity.* There exists a set A such that ∅ ∈ A and, if x ∈ A, then (x ∪ {x}) ∈ A, in which x ∪ {x} is the set x with x adjoined as a further member.

(8) *Axiom of choice.* If A is a set the elements of which are nonempty sets, then there exists a function f with domain A such that, for member B of A, f(B) ∈ B.

(9) *Axiom schema of replacement.* If A is a set and f(x, y) a formula (in which x and y are free) such that for x ∈ A there is exactly one y such that f(x, y), then there exists a set B the members of which are the y's determined by f(x, y) as x ranges over A.

(10) *Axiom of restriction (foundation axiom).* Every nonempty set A contains an element B such that A ∩ B = ∅; i.e., A and B have no elements in common.

does not allow comprehension to form a set larger than previously constructed sets. (The role of constructing larger sets is given to the power-set operation.) This leads to a situation where there is no universal set—an acceptable set must not be as large as the universe of all sets.

A very different way of avoiding Russell's paradox was proposed in 1937 by the American logician Willard Van Orman Quine. In his paper "New Foundations for Mathematical Logic," the comprehension principle allows formation of {x | ɸ(x)} only for formulas ɸ(x) that can be written in a certain form that excludes the "vicious circle" leading to the paradox. In this approach, there is a universal set.

Zermelo–Fraenkel Set Theory

Zermelo–Fraenkel set theory with the axiom of choice.

ZFC is the acronym for Zermelo–Fraenkel set theory with the axiom of choice, formulated in first-order logic. ZFC is the basic axiom system for modern (2000) set theory, regarded both as a field of mathematical research and as a foundation for ongoing mathematics. Set theory emerged from the researches of G. Cantor into the transfinite numbers and his continuum hypothesis and of R. Dedekind in his incisive analysis of natural numbers. E. Zermelo in 1908, under the influence of D. Hilbert at Göttingen, provided the first full-fledged axiomatization of set theory, from which ZFC in large part derives. Although several axiom systems were later proposed, ZFC became generally adopted by the 1960{}s because of its schematic simplicity and open-endedness in codifying the minimally necessary set existence principles needed and is now (as of 2000) regarded as the basic framework onto which further axioms can be adjoined and investigated. A modern presentation of ZFC follows.

The language of set theory is first-order logic with a binary predicat y e symbol ∈ for membership ("first-order" refers to quantification only over individuals, not e.g. properties). This language has as symbols an infinite store of variables; logical connectives (¬ for "not" , ∨ for "or", ∧ for "and", → for "implies" , and ↔ for "is equivalent to"); quantifiers (∀ for "for all" and ∃ for "there exists"); two binary predicate symbols, $=$ and ∈; and parentheses. (A more parsimonious presentation is possible, e.g. one can do with just ¬, ∨ and ∀, and leave out parentheses with a different syntax.) The formulas of the language are generated as follows: $x = y$ and $x \in y$ are (the atomic) formulas whenever x and y are variables. If φ and ψ are formulas, then so are $(\neg\varphi)$, $(\varphi\vee\psi)$, $(\varphi\wedge\psi)$, $(\varphi\rightarrow\psi)$, $(\varphi\leftrightarrow\psi)$, $\forall x\varphi$, and $\exists x\varphi$, whenever x is a variable. The various further notations can be regarded as abbreviations; for example, $x \subseteq y$ for "x is a subset of y" abbreviates $\forall z(z \in x \rightarrow z \in y)$.

The axioms of ZFC are as follows, with some historical and notational commentary.

- Axiom of extensionality:

$$\forall x\forall y\big(\forall z(z \in x \leftrightarrow z \in y) \rightarrow x = y\big).$$

This is a fundamental principle of sets, that sets are to be determined solely by their members. The arrow "can be replaced by" since the other direction is immediate. Indeed, the axiom can then be taken to be a means of introducing $=$ itself as an abbreviation, as a symbol defined in terms of ∈ . The term "extensionality" stems from a traditional philosophical distinction between the intension and the extension of a term, where loosely speaking the extension of a term is the collection of things of which the term is true of, and the intension is some more intrinsic sense of the term. A clear statement of the principle of extensionality had already appeared in the pioneering work of Dedekind, which provided a development of the natural numbers in set-theoretic terms and anticipated Zermelo's axiomatic, abstract approach to set theory.

- Axiom of the empty set:

$$\exists x \forall y (\neg y \in x).$$

This axiom asserts the existence of an empty set, such a set is unique, and is denoted by the term $\emptyset$. Terms are similarly introduced in connection with other axioms below, and in general such terms can be eliminated in favour of their definitions; for example, $\emptyset \in z$ can be regarded as an abbreviation for $\exists x (\forall y (\neg y \in x) \wedge x \in z)$.

- Axiom of pairs:

$$\forall x \forall y \exists z \forall v (v \in z \leftrightarrow (v = x \vee v = y)).$$

This axiom asserts, for any sets x and y, the existence of their (unordered) pair, the set consisting exactly of x and y. This set is denoted by $\{x,y\}$. It implies, taking its y to be x, that for any set x there is a set consisting solely of x, denoted by $\{x\}$.

The existence of $\emptyset$ and the distinction between a set x and the single-membered $\{x\}$ were not clearly delineated in the early development of set theory, and equivocations in these directions can be found:

- Axiom of union:

$$\forall x \exists z \forall v (v \in z \leftrightarrow \exists y (y \in x \wedge v \in y)).$$

This axiom asserts, for any set x, the existence of its (generalized) union, the set consisting exactly of the members of members of x. This union is denoted by $\cup x$. Note that for two sets a and b, $\cup\{a,b\}$ is the usual union $a \cup b$.

- Axiom of power set:

$$\forall x \exists z \forall v (v \in z \leftrightarrow \forall w (w \in v \rightarrow w \in x)).$$

This axiom asserts, for any set x, the existence of its power set, the set consisting exactly of those sets v that are subsets of x. This power set is denoted by $p(x)$. The axioms are generative axioms, providing various means of collecting sets together to form new sets. The generative process can be an outright existence axiom. The next axiom is another outright existence axiom, which for convenience is stated via terms defined above:

- Axiom of infinity:

$$\exists z (\emptyset \in x \wedge \forall y (y \in x \rightarrow y \cup \{y\} \in x)).$$

Among various possible approaches, this axiom asserts the existence of an infinite set of a specific kind: the set contains the empty set and is moreover closed in the sense that whenever y is in the set, so also is $y \cup \{y\}$. Hence, $\emptyset$, $\{\emptyset\}$, $\{\emptyset\{\emptyset\}\}$, $(\emptyset,\{\emptyset\},\{\emptyset,\{\emptyset,\}\})$,... are to be members; these are indeed sets and are moreover distinct from each other. Zermelo himself had $\{y\}$ in place of $y \cup \{y\}$, but the modern formulation derives from the formulation by J. von Neumann of the ordinal numbers within set theory. Dedekind had (in) famously "proved" the existence of an infinite set; Zermelo was first to see the need to postulate the existence of an infinite set. In the presence of axioms becomes a much more powerful axiom, purportly collecting together in one set all arbitrary subsets of an infinite set; Cantor famously established that no set is in bijective correspondence with its power set, and this leads to an infinite range of transfinite cardinalities.

- Axiom of choice:

$$\forall x:$$

$$\forall v \forall w \big(\big((v \in x \wedge w \in x) \wedge \exists t (t \in v \wedge t \in w) \big) \to v = w \big) \downarrow$$

$$\exists y \forall v \big((v \in x \wedge (\neg v = \varnothing)) \to \exists s \forall t ((t \in v \wedge t \in y) \leftrightarrow s = t) \big).$$

This is one of the most crucial axioms of Zermelo's axiomatization. To unravel it, the hypothesis asserts that x consists of pairwise disjoint sets, and the conclusion, that there is a set y that with each non-empty member of x has exactly one common member. Thus, y serves as a "selector" of elements from members of x) is usually stated in terms of functions: The theory of functions, construed as sets of ordered pairs with the univalent property on the second coordinate, is first developed with the previous axioms. Then axiom has an equivalent formulation as: Every set has a choice function, i.e. a function f whose domain is the set and such that for each non-empty member y of the set, $f(y) \in y$.

Zermelo formulated and with it, established his famous well-ordering theorem: Every set can be well-ordered. Zermelo maintained that the axiom of choice is a "logical principle" which "is applied without hesitation everywhere in mathematical deduction". However, Zermelo's axiom and result generated considerable criticism because of the positing of arbitrary functions following no particular rule governing the passage from argument to value. Since then, of course, the axiom has become deeply embedded in mathematics, assuming a central role in its equivalent formulation as Zorn's lemma. In response to critics, Zermelo published a second proof of his well-ordering theorem, and it was in large part to buttress this proof that he published his axiomatization, making explicit the underlying set-existence assumptions used.

- Axiom (schema) of separation: For any formula φ with unquantified variables among $v, v1 \ ,.., v_n,$

$$\forall x \forall v_1 ... \forall v_n \exists y \forall v \big(v \in y \leftrightarrow (v \in x \wedge \varphi) \big).$$

This is another crucial component of Zermelo's axiomatization. Actually, it is an infinite package of axioms, one for each formula φ, positing for any set x the existence of a subset y consisting of those members of x "separated" out according to φ. Zermelo was aware of the paradoxes of logic emerging at the time, and he himself had found the famous Russell paradox independently. Russell's paradox results from "full comprehension", the allowing of any collection of sets satisfying a property to be a set: Consider the property $(\neg y \in y)$; if there were a set R consisting exactly of those y satisfying this property, one would have the contradiction $(R \in R \leftrightarrow (\neg R \in R))$. Zermelo saw that if one only allowed collections of sets satisfying a property "and drawn from a given set" to be a set, then there are no apparent contradictions. Thus was Zermelo able to retain, in an adequate way as it has turned out, the important capability of generating sets corresponding to properties. The first theorem applies together with the Russell paradox argument to assert that the universe of sets is not itself a set.

Zermelo's version retained an intensional aspect, with his φ being some "definite" property determinate for any $y \in x$ whether the property is true of y or not. However, this became unsatisfactory in the development of set theory, and eventually the suggestion of T. Skolem of taking Zermelo's definite properties as those expressible in first-order logic was adopted, yielding. Generally speaking, logic loomed large in the formalization of mathematics at the turn into the twentieth century, at the time of G. Frege and B. Russell, but in the succeeding decades there was a steady dilution of what was considered to be logical in mathematics. Many notions came to be considered distinctly set-theoretic rather than logical, and what was retained of logic in mathematics was first-order logic.

- Axiom (schema) of replacement: For any formula φ in two unquantified variables v and w,

$$\forall v \exists u (\forall w \varphi \leftrightarrow u = w)$$
$$\downarrow \forall x \exists y \forall x (w \in y \leftrightarrow \exists v (v \in x \wedge \varphi)).$$

This also is an infinite package of axioms, one for each φ. To unravel it, the hypothesis asserts that φ is "functional" in the sense that to each set v there is a unique corresponding set u satisfying φ, and the conclusion, that for any set   there is a set y serving as the "image of x under v". In short, for any definable function correspondence and any set, the image of that set under the correspondence is also a set.

It was not part of Zermelo's original axiomatization, and to meet its inadequacies for generating certain kinds of sets, A. Fraenkel and Skolem independently proposed adjoining. Because of historical circumstance, it was Fraenkel whose initial became part of the acronym ZFC. However, it was Von Neumann's incorporation of a method into set theory, transfinite recursion, that necessitated the full exercise. In particular, he defined (what are now called the von Neumann) ordinals within set theory to correspond to Cantor's original, abstract ordinal numbers) is needed to establish that every well-ordered set is order-isomorphic to an ordinal.

- Axiom of foundation:

$$\forall x ((\neg x = \varnothing) \rightarrow \exists y (y \in x \wedge \forall z (z \in x \rightarrow \neg z \in y))).$$

This asserts that every non-empty set x is well-founded, i.e. has a "minimal" member y in terms of $\in$.

It also was not part of Zermelo's axiomatization, but appeared in his final axiomatization. It is an elegant form of the assertion that the formal universe V of sets is stratified into a cumulative hierarchy: The axiom is equivalent to the assertion that V is layered into sets V_a for (von Neumann) ordinals a, where:

$$V_0 = \varnothing; \; V_a = \bigcup_{\beta < a} p(V_{\beta+1}); \text{ and } V = \bigcup_a V_\alpha.$$

D. Mirimanoff and von Neumann had also formulated the cumulative hierarchy, but more to specific purposes. Zermelo substantially advanced the schematic generative picture with his adoption of A10), and K. Gödel urged this view of the set-theoretic universe. It is the one axiom unnecessary for the recasting of mathematics in set-theoretic terms, but the axiom is also the salient feature that distinguishes investigations specific to set theory as an autonomous field of mathematics. Indeed, it can fairly be said that current set theory is at base the study of well-foundedness, the Cantorian well-ordering doctrines adapted to the Zermelian generative conception of sets.

ZFC, again, is the standard system of axioms for set theory, given by the axioms A1)–A10) above. "Z" is the common acronym for Zermelo set theory, the axioms above but with A9), the axiom (schema) of replacement, deleted. Finally, "ZF" is the common acronym for Zermelo–Fraenkel set theory, the axiom of choice, deleted.

There has been a tremendous amount of work done in the axiomatic investigation of set theory. The first substantial result was Gödel's relative consistency result that if ZF is consistent, then so also is ZFC (and this together with Cantor's continuum hypothesis; P. Cohen in famous work leading to the Fields Medal, established the relative independence result, that if ZF is consistent, then so also is ZF together with the negation of the axiom of choice (and so also is ZFC together with the negation of the continuum hypothesis). A great deal of the work of the last several decades (as of 2000) has been devoted to the investigation of large cardinal axioms adjoined to ZFC and their consequences and interactions with ongoing mathematics.

Set Theory of the Continuum

From Cantor and until about 1940, set theory developed mostly around the study of the continuum, that is, the real line $\mathbb{R}$. The main topic was the study of the so-called regularity properties, as well as other structural properties, of simply-definable sets of real numbers, an area of mathematics that is known as Descriptive Set Theory.

Descriptive Set Theory

Descriptive Set Theory is the study of the properties and structure of definable sets of real numbers and, more generally, of definable subsets of $\mathbb{R}^n$ and other *Polish spaces* (i.e., separable, metric, and complete), such as the *Baire space* $\mathcal{N}$ of all functions $f : \mathbb{N} \to \mathbb{N}$, the space of complex numbers, Hilbert space, and separable Banach spaces. The simplest sets of real numbers are the basic open sets (i.e., the open intervals with rational endpoints), and their complements. The sets that are obtained in a countable number of steps by starting from the basic open sets and applying the operations of taking the complement and forming a countable union of previously obtained sets are the *Borel sets*. All Borel sets are *regular*, that is, they enjoy all the classical *regularity properties*. One example of a regularity property is the *Lebesgue measurability*: a set of reals is Lebesgue measurable if it differs from a Borel set by a null set, namely, a set that can be covered by sets of basic open intervals of arbitrarily-small total length. Thus, trivially, every Borel set is Lebesgue measurable, but sets more complicated than the Borel ones may not be. Other classical regularity properties are the *Baire property* (a set of reals has the Baire property if it differs from an open set by a meager set, namely, a set that is a countable union of sets that are not dense in any interval), and the *perfect set property* (a set of reals has the perfect set property if it is either countable or contains a perfect set, namely, a closed set with no isolated points). In ZFC one can prove that there exist non-regular sets of reals, but the AC is necessary for this (Solovay 1970).

The *analytic sets*, also called $\sum_1^1$, are the continuous images of Borel sets. And the *co-analytic*, or $\prod_1^1$, sets are the complements of analytic sets.

Starting from the analytic (or the co-analytic) sets and applying the operations of projection (from the product space $\mathbb{R} \times \mathcal{N}$ to $\mathbb{R}$) and complementation, one obtains the *projective sets*. The projective sets form a hierarchy of increasing complexity. For example, if $A \subseteq \mathbb{R} \times \mathcal{N}$ is co-analytic, then the projection $\{x \in \mathbb{R} : \exists y \in \mathcal{N}((x,y) \in A)\}$ is a projective set in the next level of complexity above the co-analytic sets. Those sets are called $\sum_2^1$, and their complements are called $\prod_2^1$.

The projective sets come up very naturally in mathematical practice, for it turns out that a set AA of reals is projective if and only if it is definable in the structure

$$R = (\mathbb{R}, +, \cdot, \mathbb{Z}).$$

That is, there is a first-order formula $\varphi(x, y_1, \ldots, y_n)$ in the language for the structure such that for some $r_1, \ldots, r_n \in \mathbb{R}$,

$$A = \{x \in \mathbb{R} : R \vDash \varphi(x, r_1, \ldots, r_n)\}.$$

ZFC proves that every analytic set, and therefore every co-analytic set, is Lebesgue measurable and has the Baire property. It also proves that every analytic set has the perfect set property. But the perfect set property for co-analytic sets implies that the first uncountable cardinal, $\aleph_1$, is a large cardinal in the constructible universe LL, namely a so-called inaccessible cardinal, which implies that one cannot prove in ZFC that every co-analytic set has the perfect set property.

The theory of projective sets of complexity greater than co-analytic is completely undetermined by ZFC. For example, in LL there is a $\sum_2^1$ set that is not Lebesgue measurable and does not have the Baire property, whereas if Martin's axiom holds every such set has those regularity properties. There is, however, an axiom, called the axiom of Projective Determinacy, or PD, that is consistent with ZFC, modulo the consistency of some large cardinals, and implies that all projective sets are regular. Moreover, PD settles essentially all questions about the projective sets.

Determinacy

A regularity property of sets that subsumes all other classical regularity properties is that of being *determined*. For simplicity, we shall work with the Baire space $\mathcal{N}$. Recall that the elements of $\mathcal{N}$ are functions $f : \mathbb{N} \to \mathbb{N}$, that is, sequences of natural numbers of length ω. The space $\mathcal{N}$ is topologically equivalent (i.e., homeomorphic) to the set of irrational points of $\mathbb{R}$. So, since we are interested in the regularity properties of subsets of $\mathbb{R}$, and since countable sets, such as the set of rationals, are negligible in terms of those properties, we may as well work with $\mathcal{N}$, instead of $\mathbb{R}$.

Given $A \subseteq \mathcal{N}$, the *game* associated to A, denoted by $\mathcal{G}_A$, has two players, I and II, who play alternatively $n_i \in \mathbb{N}$ I plays n_0, then II plays n_1, then I plays n_2, and so on. So, at stage 2k, player I plays n_{2k} and at stage $2k+1$, player II plays n_{2k+1}. We may visualize a run of the game as follows:

I	n_0	n_2	n_4	...	n_{2k}	...
II		n_1	n_3	...	...	n_{2k+1} ...

After infinitely many moves, the two players produce an infinite sequence $n_0, n_1, n_2, \ldots$ of natural numbers. Player I wins the game if the sequence belongs to AA. Otherwise, player II wins.

The game $\mathcal{G}_A$ is *determined* if there is a winning strategy for one of the players. A *winning strate-gy*for one of the players, say for player II, is a function σσ from the set of finite sequences of natural numbers into $\mathbb{N}$, such that if the player plays according to this function, i.e., she plays $\sigma(n_0, \ldots, n_{2k})$ at the kk-th turn, she will always win the game, no matter what the other player does.

We say that a subset AA of $\mathcal{N}$ is *determined* if and only if the game $\mathcal{G}_A$ is determined.

One can prove in ZFC—and the use of the AC is necessary—that there are non-determined sets. Thus, the *Axiom of Determinacy* (AD), which asserts that all subsets of $\mathcal{N}$ are determined, is incompatible with the AC. But Donald Martin proved, in ZFC, that every Borel set is determined. Further, he showed that if there exists a large cardinal called measurable then even the analytic sets are determined. The axiom of *Projective Determinacy* (PD) asserts that every projective set is determined. It turns out that PD implies that all projective sets of reals are regular, and Woodin has shown that, in a certain sense, PD settles essentially all questions about the projective sets. Moreover, PD seems to be necessary for this. Another axiom, $AD^{L(\mathbb{R})}$, asserts that the AD holds in $L(\mathbb{R})$, which is the least transitive class that contains all the ordinals and all the real numbers, and satisfies the ZF axioms So, $AD^{L(\mathbb{R})}$ implies that every set of reals that belongs to $L(\mathbb{R})$ is regular. Also, since $L(\mathbb{R})$ contains all projective sets, $AD^{L(\mathbb{R})}$ implies PD.

The Continuum Hypothesis

The Continuum Hypothesis (CH), formulated by Cantor in 1878, asserts that every infinite set of real numbers has cardinality either $\aleph_0$ or the same cardinality as $\mathbb{R}$. Thus, the CH is equivalent to $2^{\aleph_0} = \aleph_1$.

Cantor proved in 1883 that closed sets of real numbers have the perfect set property, from which it follows that every uncountable closed set of real numbers has the same cardinality as RR. Thus, the CH holds for closed sets. More than thirty years later, Pavel Aleksandrov extended the result to all Borel sets, and then Mikhail Suslin to all analytic sets. Thus, all analytic sets satisfy the CH. However, the efforts to prove that co-analytic sets satisfy the CH would not succeed, as this is not provable in ZFC.

In 1938 Gödel proved the consistency of the CH with ZFC. Assuming that ZF is consistent, he built a model of ZFC, known as the *constructible universe*, in which the CH holds. Thus, the proof shows that if ZF is consistent, then so is ZF together with the AC and the CH. Hence, assuming ZF is consistent, the AC cannot be disproved in ZF and the CH cannot be disproved in ZFC.

Countable and Uncountable Sets

A set A is said to be finite, if A is empty or there is n ∈ N and there is a bijection f : {1,... ,n} → A. Otherwise the set A is called infinite. Two sets A and B are called equinumerous, written A ∼ B, if there is a bijection f : X → Y. A set A is called countably infinite if $A \sim \mathbb{N}$. We say that A is countable if $A \sim \mathbb{N}$ or A is finite.

Example: The sets $(0,\infty)$ and R are equinumerous. Indeed, the function $f : \mathbb{R} \to (0,\infty)$ defined by $f(x) = e^x$ is a bijection.

Example: The set $\mathbb{Z}$ of integers is countably infinite. Define $f : \mathbb{N} = \to \mathbb{Z}$ by

$$f(n) = \begin{cases} n/2 & \text{if } n \text{ is even}; \\ -(n-1)/2 & \text{if } n \text{ is odd}. \end{cases}$$

Then f is a bijection from $\mathbb{N}$ to $\mathbb{Z}$ so that $\mathbb{N} \sim \mathbb{Z}$.

If there is no bijection between $\mathbb{N}$ and A, then A is called uncountable.

Theorem: There is no surjection from a set A to $P(A)$.

Proof: Consider any function $f : A \to P(A)$ and let

$$B = \{a \in A \mid a \notin f(a)\}.$$

We claim that there is no $b \in A$ such that f(b) = B. Indeed, assume f(b) = B for some $b \in A$. Then either $b \in B$ hence $b \notin f(b)$ which is a contradiction, or $b \notin B = f(b)$ implying that $b \in B$ which is again a contradiction. Hence the map f is not surjective as claimed.

As a corollary we have the following result.

Corollary: The set $P(\mathbb{N})$ is uncountable.

Proposition: Any subset of a countable set is countable.

Proof: Without loss of generality we may assume that A is an infinite subset of $\mathbb{N}$. We define $h : \mathbb{N} \to A$ as follows. Let h(1) = min A. Since A is infinite, A is nonempty and so h() is well-defined. Having defined $h(n-1)$, we define $h(n) = min(A \setminus \{h(1),\dots,h(n-1)\})$. Again since A is infinite the set $(A \setminus \{h(1),\dots,h(n-1)\})$ is nonempty, h(n) is well-defined. We claim that h is a bijection. We first show that h is an injection. To see this we prove that $h(n+k) > h(n)$ for all $n, k \in \mathbb{N}$. By construction $h(n+1) > h(n)$ for all $n \in \mathbb{N}$. Then setting $B = \{k \in \mathbb{N} \mid h(n+k) > h(n)\}$ we see that $1 \in B$ and if $h(n + (k-1)) > h(n)$, then $h(n+k) > h(n + (k-1)) > h(n)$. Consequently, $B = \mathbb{N}$. Since n was arbitrary, $h(n+k) > h(n)$ for all $n, k \in \mathbb{N}$. Now taking distinct $n, m \in \mathbb{N}$ we may assume that m > n so that m = n + k. By the above h(m) = h(n + k) > h(n) proving that h is an injection. Next we show that h is a surjection. To do this we first show that $h(n) \geq n$. Let $C = \{n \in \mathbb{N} \mid h(n) \geq n\}$. Clearly, $1 \in C$. If $k \in C$, then h(k + 1) > h(k) $\geq$ n so that h(k + 1) $\geq$ k + 1. Hence k + 1 $\in$ C and by the principle of mathematical induction $C = \mathbb{N}$. Now take $n_0 \in A$. We have to show that $h(m_0) = n_0$ for some $m_0 \in \mathbb{N}$. If $n_0 = 1$, then $m_0 = 1$. So assume that $n_0 \geq 2$. Consider the set $D = \{n \in A \mid h(n) \geq n_0\}$. Since $h(n_0) \geq n_0$, the set D is nonempty and by the well-ordering principle D has a minimum. Let $m_0 = min D$. If $m_0 = 1$, then $h(m_0) = min A \leq n_0 \leq h(m_0)$ and hence $h(m_0) = n_0$. So we may also assume that n> min A. Then $h(m_0) \geq n_0 > h(m_0 - 1) > \dots > h(1)$ in view of definitions of m_0 and h. Since $h(m_0) = min(A \setminus \{h(1),\dots,h(m_0 - 1)\})$ and $n_0 \in A \setminus \{h(1),\dots,h(m_0 - 1)\}$ and $h(m_0) \geq n_0$, it follows that $h(m_0) = n_0$. This proves that h is also a surjection.

Proposition: Let A be a non-empty set. Then the following are equivalent.

(a) A is countable.

(b) There exists a surjection $f : \mathbb{N} \to A$.

(c) There exists an injection $g : A \to \mathbb{N}$.

Proof: $(a) \implies (b)$ If A is countably infinite, then there exists a bijection $f : \mathbb{N} \to A$ and then (b) follows. If A is finite, then there is bijection $h : \{1,\dots,n\} \to A$ for some n. Then the function $f : \mathbb{N} \to A$ defined by

$$f(i) = \begin{cases} h(i) & 1 \le i \le n, \\ h(n) & i > n. \end{cases}$$

is a surjection.

$(b) \implies (c)$. Assume that $f : \mathbb{N} \to A$ is a surjection. We claim that there is an injection $g : A \to \mathbb{N}$. To define g note that if $a \in A$, then $f^{-1}(\{a\}) \ne \varnothing$. Hence we set $g(a) = \min f^{-1}((a)\})$. Now note that if $a \ne a'$, then the sets $f^{-1}(\{a\}) \cap f^{-1}(\{a'\}) = \varnothing$ which implies $\min^{-1}(\{a\}) \ne \min^{-1}(\{a'\})$. Hence $g(a) \ne g(a')a$ and $g : A \to \mathbb{N}$ is an injective. (c) $\Rightarrow$ (a). Assume that $g : A \to \mathbb{N}$ is an injection. We want to show that A is countable. Since $g : A \to g(A)$ is a bijection and $g(A) \subset N$, Proposition implies that A is countable.

Corollary: The set $\mathbb{N} \times \mathbb{N}$ is countable.

Proof: By Proposition 3.6 it suffices to construct an injective function $f : \mathbb{N} \times \mathbb{N} \to \mathbb{N}$. Define $f : \mathbb{N} \times \mathbb{N} \to \mathbb{N}$ by $f(n,m) = 2^n\, 3^m$. Assume that $2^n\, 3^m = 2^k\, 3^l$. If n < k, then $3^m = 2^{k-n}\, 3^l$. The left side of this equality is an odd number whereas the right is an even number implying $n = k$ and $3^m = 3^l$. Then also $m = l$. Hence f is injective.

Proposition: If A and B are countable, then $A \times B$ is countable.

Proof: Since A and B are countable, there exist surjective functions $f : \mathbb{N} \to A$ and $g : \mathbb{N} \to B$. Define $\mathbb{N} \times \mathbb{N}$ by $F(n,m) = (f(n), g(m))$. The function F is surjective. Since $\mathbb{N} \times \mathbb{N}$ is countably infinite, there is a bijection $h : N \to N \times N$. Then $G : \mathbb{N} \times A \times B$ defined by G = F $\circ$ h is a surjection. By part (c) of Proposition, the set $A \times B A \times B$ is countable.

Corollary: The set $\mathbb{Q}$ of all rational numbers is countable.

Proposition: Assume that the set I is countable and A_i is countable for every $i \in I$. Then $\bigcup_{i \in I} A_i$ is countable.

Proof: For each $i \in I$, there exists a surjection $f_i : \mathbb{N} \to A_i$. Moreover, since S I is countable, there exists a surjection $g : \mathbb{N} \to I$. Now define $h : \mathbb{N} \times \mathbb{N} \to \bigcup_{i \in I} A_i$ by $F(n,m) = f_{g(n)}(m)$ and let $h : \mathbb{N} \to \mathbb{N} \times \mathbb{N}$ be a bijection. Then F is a surjection and the composition $G = F \circ h : \mathbb{N} \to \bigcup_{i \in I} A_i$ is a surjection. By Proposition, $\bigcup_{i \in I} A_i$ is countable.

Proposition: The set of real numbers $\mathbb{R}$ is uncountable.

The proof will be a consequence of the following result about nested intervals.

Proposition: Assume that $(In)_{n \in \mathbb{N}}$ is a countable collection of closed and bounded intervals $I_n = a_n, b_n]$ satisfying $I_{n+1} \subset I_n$ for all $n \in \mathbb{N}$. Then $\bigcap_{n \in \mathbb{N}} I_n \neq \emptyset$.

Proof: Since $[a_{n+1}, b_{n+1}] \subset [a_n, b_n]$ for all n, it follows that $a_n \leq b_k$ for all $n, k \in \mathbb{N}$. So, the set $A = \{a_n \mid n \in \mathbb{N}\}$ is bounded above by every b_k and consequently $a := \sup A \leq b_k$ for all $k \in \mathbb{N}$. But this implies that the set $B = \{b_k \mid k \in \mathbb{N}\}$ is bounded below by a so that $a \leq b := \inf B$. Hence $\bigcap_{n \in \mathbb{N}} I_n = [a, b]$.

Proof of Proposition: Arguing by contradiction assume that $\mathbb{R}$ is countable. Let $x_1, x_2, x_3, ...$ be enumeration of $\mathbb{R}$. Choose a closed bounded interval I_1 such that $x_1 \notin I_1$. Having chosen the closed intervals $I_1, I_2, ... , I_{n-1}$, we choose the closed interval I_n to be a subset of I_{n-1} such that $x_n \notin I_n$. Consequently, we have a countable collection of closed bounded intervals (I_n) such that $I_{n+1} \subset I_n$ and $x_n \notin I_n$. Then by the above proposition, $\bigcap_{n \in \mathbb{N}} I_n \neq \emptyset$. Observe that if x belongs to this intersection, then x is not on the list $x_1, x_2, ...$, contradiction.

Cantor's Theorem

For any set X , the *power set* of X (*i.e.*, the set of subsets of X), is larger (has a greater cardinality) than X .

Cantor's Theorem tells us that no matter how large a set we have, we may consider a set that is still larger. This is trivial if the set in question has finitely many members, but not at all obvious if our set is infinite.

The word larger in this usage has a precise meaning. In set theory two sets are considered to be the same 'size' if we can form a one-to-onematch-up between the elements of the two sets with none left over on either side. (This is called a bijection in mathematics.) If the sets are finite, this means they have the same number of elements. However, if the sets are infinite then it isn't quite clear how the ordinary concept of 'number of elements' should apply, and so we need this notion of one-to-one match-up to make it perfectly clear.

For example, if you have more pennies than dimes and you attempt to form a one-to-one matchup between them, you will find that you always have a penny left over:

One-to-one match-up.

Cantor realized that the same principle can be applied to infinite sets, and discovered that no matter what set you start with, any attempt to form a one-to-one match-up of the elements of the set to the subsets of the set *must* leave some subset unmatched.

The proof uses a technique that Cantor originated called diagonalization, which is a form of proof by contradiction. We begin by assuming that for some set we can form a one-to-one match-up between the elements and the subsets that leaves none out on either side. Let's call the set X, and we'll denote the power set by $\mathcal{P}(X)$:

$$X = \{a,b,c,d,e,f,...\}$$

and

$$P(X) = \{\{a\},\{b\},\{a,b\},\{c\},\{a,c\},$$
$$\{b,c\},\{a,b,c\},\{d\},\{a,d\}...\}$$

Now the one-to-one matchup (that we are *assuming* exists) between the elements and the power sets might look something like this:

$$X \begin{cases} a & \to & \{c,d\} \\ b & \to & \{a\} \\ c & \to & \{a,b,c,d\} \\ d & \to & \{b,e\} \\ e & \to & \{a,c,e\} \\ \vdots & & \vdots \end{cases} PX$$

Notice that some of the elements of X are matched to subsets that contain them. For example, in the hypothetical match-up pictured above the element ee is matched to the subset $\{a,c,e\}$. Other elements might match to subsets that *do not* contain them. For example, above the element aa is matched with the subset $\{c,d\}$. For any given match-up we may define the subset of X consisting of just those elements that are matched to subsets that do not contain them. By the assumption that our match-up exists this is a well-defined subset. Call it F.

This F, being a subset of X, has to appear somewhere in the match-up, with some element being matched to it. But what element could it be? Whatever element it is, it can't be ssmatch to subsets that contain them. On the other hand, if it isn't an element of F—why then it must be, since it would then match to a subset that doesn't contain it, forcing it to be a member of F by definition.

The only possible conclusion is that no element can be matched to F, which means that our one-to-one match-up can't be complete after all.

You might need to read that again. Then take a minute to ponder what it means. Cantor's Theorem proves that given any set, even an infinite one, the set of its subsets is 'bigger' in a very precise sense. But this means there can be no largest infinity either. The kinds of infinity are (ahem) infinite.

References

- Set-theory: geeksforgeeks.org, Retrieved 21 May 2018

- Operations-on-sets: math-only-math.com, Retrieved 30 June 2018

- Intersection-of-sets: math-only-math.com, Retrieved 31 March 2018

- Axiomatic-set-theory, set-theory, science: britannica.com, Retrieved 18 May 2018

- Russells-paradox: britannica.com, Retrieved 31 May 2018

- Cantors-Theorem, encyclopedia: platonicrealms.com, Retrieved 22 June 2018

Permissions

Index